Laboratories of Art

Archimedes

NEW STUDIES IN THE HISTORY AND PHILOSOPHY OF SCIENCE AND TECHNOLOGY

VOLUME 37

Archimedes has three fundamental goals; to further the integration of the histories of science and technology with one another: to investigate the technical, social and practical histories of specific developments in science and technology; and finally, where possible and desirable, to bring the histories of science and technology into closer contact with the philosophy of science. To these ends, each volume will have its own theme and title and will be planned by one or more members of the Advisory Board in consultation with the editor. Although the volumes have specific themes, the series itself will not be limited to one or even to a few particular areas. Its subjects include any of the sciences, ranging from biology through physics, all aspects of technology, broadly construed, as well as historically-engaged philosophy of science or technology. Taken as a whole, *Archimedes* will be of interest to historians, philosophers, and scientists, as well as to those in business and industry who seek to understand how science and industry have come to be so strongly linked.

For further volumes:
http://www.springer.com/series/5644

Sven Dupré
Editor

Laboratories of Art

Alchemy and Art Technology from Antiquity to the 18th Century

Springer

Editor
Sven Dupré
Freie Universität
Max Planck Institute for the History of Science
Berlin, Germany

ISSN 1385-0180 ISSN 2215-0064 (electronic)
ISBN 978-3-319-05064-5 ISBN 978-3-319-05065-2 (eBook)
DOI 10.1007/978-3-319-05065-2
Springer Cham Heidelberg New York Dordrecht London

Library of Congress Control Number: 2014937822

Printed on acid-free paper

Springer is part of Springer Science+Business Media (www.springer.com)

Contents

Introduction

The Antwerp painter Adriaen van Utrecht (1599–1652) arranged Chinese porcelain, a goblet made of rock crystal and *cristallo* glasses *à la façon de Venise*, next to a magnificent display of gold- and silversmiths' works on a table in the foreground of his 1636 "Allegory of Fire" (Fig. 1).[1] Many of these objects have been identified as originating in Antwerp, and it is likely that Van Utrecht's painting celebrates the manufacture and trade of luxury goods for which Antwerp gained fame in the early seventeenth century. Van Utrecht's objects have another point in common: one way or the other they are produced through the agency of fire. Van Utrecht's painting celebrates Antwerp's economic power and the productive ingenuity of its craftsmen and artists as much as it praises the transformative force of fire turning relatively cheap and humble materials into highly valued objects of art. These objects were all products of the arts of fire, which according to Vannoccio Biringuccio (1480–c.1539) included alchemy, and excluded "false alchemy" concerned with pretentious transmutation.[2]

In the background of Van Utrecht's painting, a window opens on to a space in which a man stands working at a stove with an open fire. Stirring a cauldron, the man is shown in the material company of bellows, an anvil, a melting and a distilling furnace, and other equipment related to the worlds of assaying and metallurgy. Van Utrecht's background refers to the spaces in which the objects in the foreground were produced. Were these spaces laboratories or artisanal workshops? Were they home to gold- and silversmiths, glassmakers or producers of porcelain? The difference between workshops and laboratories during this period is ambiguous, not only in Van Utrecht's depiction, and the line between the two spaces is as difficult to draw as that between the arts of fire and alchemy. This book is concerned with the interconnections and differentiations between foreground and

[1] My discussion of Van Utrecht's painting in this and the next paragraph is based on Göttler, "The Alchemist, the Painter."

[2] For Biringuccio's attitude towards alchemy, see Newman, *Promethean Ambitions,* 128–32.

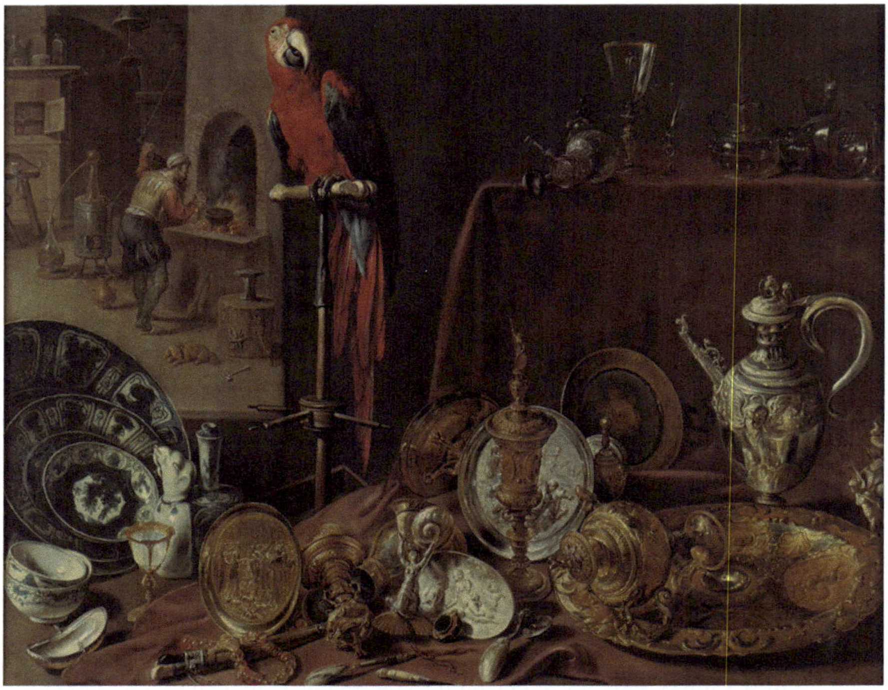

Fig. 1 Adriaen van Utrecht, *Allegory of Fire*, 1636 (Courtesy of Royal Museums of Fine Arts of Belgium, Brussels (Photo: J. Geleyns))

background in Van Utrecht's painting, between artisanal workshops and alchemical laboratories, between the material arts and alchemy.

Van Utrecht is one of the heirs of Van Eyck (c.1390–1441). Following Giorgio Vasari (1511–1574), Karel van Mander (1548–1606) portrayed Van Eyck as an alchemist and likened his invention of oil paint to alchemical experimentation.[3] According to Van Mander, Van Eyck experimented with varnishes and binding agents, "and found after much investigation that pigments mixed with such oils became malleable and dried hard, and having dried became impermeable, and that the oil made colours livelier, and that they themselves became lustrous without varnishing."[4] It turns out that Van Eyck did not invent oil paint, and neither was he an alchemist.

Nevertheless he would never have been described in those terms had not artists and alchemists both been considered agents of material transformation. Van Mander's description of Van Eyck's experimentation also reminds us that painters and alchemists shared materials, and that painters used artificially created pigments, "made by alchemy" according to Cennino Cennini (c.1370–c.1440). However, this

[3] Davis, "Renaissance Inventions."

[4] Mander, *Het schilder-boeck*, 199v. Translation in Melion, *Shaping the Netherlandish Canon*, 79.

book is not concerned with travelling materials and shared material culture. Shifting the focus from painting to the decorative arts, this book scrutinizes epistemic exchanges between producers of the arts of fire and alchemists.

Laboratories and Workshops

What can the evolution of the laboratory, and its shifting relation to the artisanal workshop, tell us about epistemic exchanges between the arts and alchemy? In the fifteenth and sixteenth centuries, the term *laboratorium* uniquely referred to workplaces in which "chemical" operations were performed: smelting, combustion, distillation, dissolution, and precipitation. Matteo Martelli has convincingly shown that no such term was available in Antiquity.[5] The Papyri, containing recipes for the production of a range of luxury goods, and the earliest texts on alchemy, used the terminology of specific crafts (dyeing, metalworking and glassmaking). While artisans labored in workshops with tools and ingredients specific to their craft, alchemists seemed to have made use of a more complex set of materials, instruments and techniques belonging to various crafts. However, Martelli also shows that even the earliest alchemists developed specialized equipment. This leads to the assumption that as early as the first century AD, alchemists in Hellenistic Egypt began to differentiate themselves from artisans. However, there was no workplace specifically designed for alchemists to engage in their activities. It is likely they accessed the material equipment in artisanal workshops.

By the early modern period, the alchemist had acquired a specific place of work. According to Pamela Smith, by the mid-sixteenth century, these laboratories were referred to formally as *laboratorium* and *officina*.[6] The ubiquity of furnaces and the use of fire demanded a specific workplace, whereas experimental philosophy still lacked specific places of experimentation in the seventeenth century. Laboratories were associated with secretive practices. When in the early seventeenth century Andreas Libavius described the ideal workplace of the chemist, he emphasized how it differed from the dark, smelly, secretive laboratory that was noticeably lacking in decorum:

> We do not want the chemist to neglect the exercises of piety or exempt himself from other duties of an upright life, simply pining away amidst his dark furnaces [. . .]. Thus we are not going to devise from him just a *chymeion* or laboratory to use as a private study and hideaway in order that his practice will be more distinguished than anyone else's; but rather, what we shall provide for him is a dwelling suitable for decorous participation in society and living the life of a free man [. . .].[7]

[5] Martelli, "Greek Alchemists at Work."

[6] Smith, "Laboratories," 299.

[7] Libavius, "Commentariorum alchymiae." Quoted and translated in Hannaway, "Laboratory Design," 599. However, for corrections of Hannaway's view, see Shackelford, "Tycho Brahe, Laboratory Design," and Newman, "Chemical House of Libavius."

While much has been made of this association with secrecy to dissociate experimental philosophy from the alchemist's laboratory, Ursula Klein has revealed a continuing laboratory tradition reaching into the eighteenth century.[8] According to Klein, in contrast to experimental philosophy, work in the laboratory was characterized by continuous experimentation with material substances. The material culture supporting this laboratory work remained largely unchanged. The exception is the material equipment for transmutational alchemy and the making of the Philosophers' Stone, which generally disappeared from laboratories together with transmutational alchemy itself in the eighteenth century. Laboratories were places in which the study of nature and technical innovation went hand in hand.

As already mentioned, the term "laboratory" was not only used to refer to the workplace of the alchemist. All workplaces in which "chemical" operations were performed were known as *laboratories*, and as such, artisanal workshops with furnaces and fire in which "chemical" operations were performed were also known as laboratories. For example, the places in which sixteenth-century producers of fireworks developed their materials were called laboratories.[9] In terms of material culture the *laboratorium*, for example, one designed and equipped by Count Wolfgang II von Hohenlohe-Weidersheim (1546–1610) in the early seventeenth century, was remarkably similar to the workplaces of apothecaries, metalworkers and glassmakers.[10]

Before the emergence of mercantilist states, Renaissance courts were the most important agents in establishing laboratories. The laboratories of Rudolf II (1552–1612) in Prague and Landgrave Moritz von Hessen-Kassel (1572–1632) were thought to bring material and intellectual rewards.[11] Moritz even founded a chair of *chymiatria* at the University of Marburg in 1609.[12] The teaching of Johannes Hartmann (1561–1631), who was appointed to the chair, included work in the laboratory, likewise established at the university. Courts also developed workplaces in which the work of artisans could be tested. Most famously, in the 1670s, Johann Joachim Becher (1635–1682) proposed the establishment of a *Kunst- und Werckhaus* at the court in Vienna.[13] It would have included several different sorts of manufactures: porcelain making, silk and wool weaving, the production of medicines and glassmaking. Chemical laboratories were to form the core of the

[8] Klein, "Laboratory Challenge," and "Apothecary Shops." For the dissociation of experimental philosophy from the alchemical laboratory, see Shapin, "House of Experiment."

[9] Werrett, *Fireworks*, 29–30.

[10] Weyer, *Graf Wolfgang II. von Hohenlohe*. For material culture of laboratories, see also Osten, *Alchemistenlaboratorium Oberstockstall*; Soukup & Mayer, *Alchemistisches Gold*, and the shorter report, Soukop, Osten & Mayer, "Alembics, Cucurbits, Phials," 25; Howard, *La Bibliothèque et le laboratoire*; Pereira, "Utili segreti"; Rouaze, "Atelier de distillation"; and Joly, "Laboratoire alchimique."

[11] Among the numerous publications on court alchemy, see two foundational studies: Moran, *The Alchemical World*; and Evans, *Rudolf II*.

[12] Ganzenmüller, "Das chemische Laboratorium."

[13] Smith, *Business of Alchemy*, 190–8.

house. These laboratories were intended to specialize in salts and acids used in the production of mineral dyes. Becher envisioned that the manufacture and export of these dyes would finance the other operations of the *Kunst- und Werckhaus*. He also envisaged the house attracting artisans from various crafts and that it would function as a deposit of recipes and descriptions of techniques. Trained artisans would then be sent out to distribute new processes, new manufactures and inventions. Becher used alchemy to intrigue his patron, who was always interested in the wealth promised by metallic transmutation, and to link the worlds of the court and commerce.

Such court projects often took inspiration from the two earliest examples of spaces bringing together alchemy and the decorative arts at the Medici court in Florence. Two essays in this volume deal with these Florentine workplaces: Fanny Kieffer discusses the *Uffizi*, and Marco Beretta the *Casino di San Marco*. Founded in 1586 by Francesco I de' Medici (1541–1587), and developed by his successor, Ferdinando I (1549–1609), the *Uffizi* housed the *fonderia*, a workspace for the arts of fire. Here medicines were prepared and metals fused. The *fonderia* was simultaneously a pharmacy, an alchemical laboratory, a smithy, a goldsmith's workshop, and a confectionary. The activities performed in the *Uffizi* included everything from preparing jams and sugar sculptures to the production of glass and fireworks (and less peaceful applications of saltpeter). Francesco I also established the *Casino di San Marco*. It included an alchemical laboratory, a furnace for producing porcelain, and a glass workshop in which, under the direction of Niccolo Sisti, *cristallo* (so named because it was as clear as rock crystal) was made, as well as all sorts of colored glass in imitation of precious stones. Other important activities in the *Casino* were the fusion of rock crystal, the counterfeiting of precious stones, the production of fireworks, and the preparation of pharmaceutical remedies. Similar activities were developed in the *Uffizi* and the *Casino*, and recipes and personnel travelled between the two places. The only exception was Medici porcelain, which was exclusively produced in the workshops of the *Casino*.

In sum, Renaissance courts established spaces where artisanal workshops and laboratories were brought together facilitating the circulation of materials, people and knowledge between the worlds of craft (today's decorative arts) and alchemy.

Art Technologies and Knowledge of Material Transformation

Laboratories were not only the workplaces of transmutational alchemists. The *Uffizi* and the *Casino* had little in common with the workplaces of alleged goldmakers destined to fail and bring their families to financial ruin, as famously depicted by Pieter Brueghel the Elder (c.1525–1569) (Fig. 2). Just as the spaces known as laboratories encompassed the workplaces of artisans laboring at furnaces and

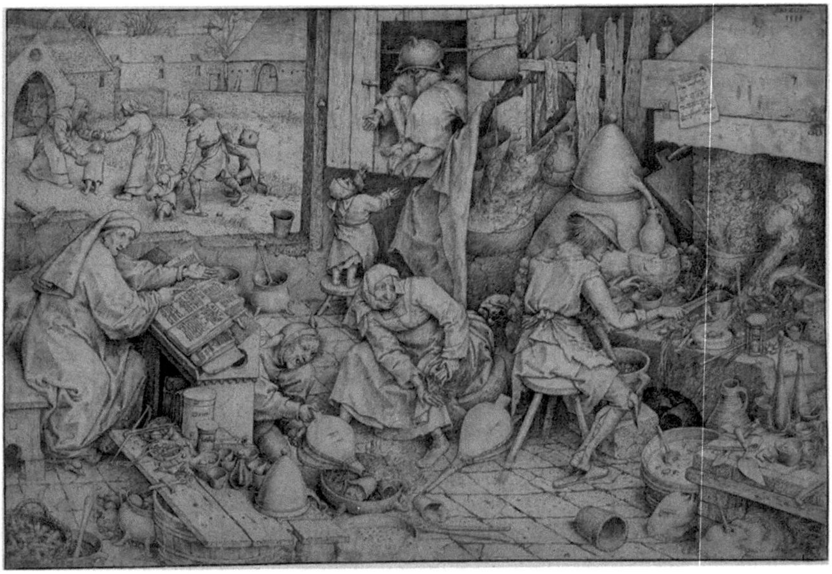

Fig. 2 Pieter Brueghel the Elder, *The Alchemist's Family: Al-gemist*, 1558 (Courtesy of the Kupferstichkabinett, Staatliche Museen zu Berlin)

using materials and chemical procedures to make both everyday items and objects of the visual and decorative arts, the activity known as alchemy encompassed more than attempts to make gold.[14] Transmutational alchemy was about the transmutation of all base metals into more noble ones, but chrysopoeia was only one aspect of alchemy. Alchemy also touched on medicine and chemical manufacture. It was about the chemical production of things—medicines, porcelain, dyes, and other products as well as the precious metals—and about the knowledge of how to produce them. In this sense, "art technologies"—materials and techniques to make art and knowledge of these materials and techniques—overlapped with alchemy.

Alchemy has deep roots in writings on material transformation from Antiquity. The productive knowledge associated with material transformation was written down in recipe books. The Leiden and Stockholm Papyri date to the third century AD. They contain recipes for the making of gold and silver, for the imitation of precious stones, and for textile dyes. Whether a recipe is about coloring silver to make it look like gold or the making of an imitation ruby or another artificial precious stone, all of the recipes in the Papyri are, indeed, about material transformation and color change as related to the manufacture of luxury goods. As Matteo Martelli shows in this volume, relatively early on a historical process of selection, appropriation and differentiation resulted in a more limited definition of

---

[14] For the scope of early modern alchemy and the notion of 'chymistry', see Principe, *Secrets of Alchemy*; and Newman & Principe, "Alchemy vs. Chemistry."

alchemy primarily focused on the making of gold and silver. Already around 300 AD, a distinction was introduced between a limited definition of alchemy as metallic transmutation and a more encompassing definition including productive knowledge and various artisanal technologies.

Transmitted to Europe, and translated into Latin and the vernaculars, several of the recipes in the Papyri are still found in collections of recipes in the fifteenth and sixteenth centuries. More importantly, as Martelli remarks in his essay, the scope of a recipe collection such as the *Mappae clavicula*, compiled between the ninth and the twelfth centuries, is as encompassing as that of the Papyri despite the earlier attempts to limit alchemy to metallic transmutation only. It should not come as a surprise then that Sylvie Neven finds it difficult to demarcate between alchemical and art technological recipes in late medieval collections of recipes. Her contribution to this volume shows that alchemical and art technological recipes shared a concern with the same materials and artisanal processes. The focus on chromatic transformation already found in the Papyri continues in fifteenth-century recipes. Although related to laboratory practices, Neven emphasizes that these recipe collections are the products of scribal compilation and copying processes. Words and works are equally important elements of alchemical practices.[15] Next to laboratories, medieval religious institutions were also important sites of alchemical and art technological practice. However, as Neven points out, this does not exclude that some scribes, such as the Benedictine monk Wolfgang Seidel (1491–1562), tried out recipes.

The scope of alchemy was from its very beginnings contested, and remained so throughout its long history. Especially its boundaries with art technology were fluid. Throughout the Middle Ages and the Renaissance, the distinction between alchemy as strictly goldmaking and a more encompassing definition overlapping with art technology was crucial to the polemics of artists and alchemists and the rivalry between alchemy and the arts.[16] Given the contested nature of the field of inquiry, it follows that the identity of the alchemist was equally contested and complex. In the early modern period the alchemist was often portrayed as a fraud. The portrayals of laboratory scenes building on Brueghel became a genre of its own in the Netherlands.[17] In these scenes, the alchemist is a goldmaker searching in vain for the Philosophers' Stone and riches. As Tara Nummedal has convincingly shown, the portrayal of the alchemist as a fraud also created the opportunity for other alchemists to fashion themselves in the role of experts, offering their services to courts and other patrons to help them unmask fraudulent alchemists.[18] Alchemical expertise was based on a broader knowledge of matter and materials.

[15] For reading and writing as alchemical practices, see Nummedal, "Words and Works," and the literature quoted there.

[16] Newman, *Promethean Ambitions*, 115–63.

[17] Principe & DeWitt, *Transmutations*, 11–27.

[18] For experts and entrepreneurial alchemy, see Nummedal, *Alchemy and Authority*, 40–5.

It is in this context of alchemical expertise that we should consider the artisans who adopted the language of alchemy in the early modern period to attract the attention of patrons. One example can be seen in the description of Black Berthold, the legendary discoverer of gunpowder in Europe, in the late fifteenth-century manuscript *Feuerwerckbuch*. Depicted in the company of a furnace and alembics, Berthold's creation of ordnance is identified with alchemy. Black Berthold, the *Feuerwerckbuch* reads,

> worked with the great alchemy like those masters who are engaged with precious and valuable things, with silver and gold, and with similar metals. These masters can separate silver and gold from other precious jewelry, and from other valuable colors which they can produce. Now this master Berthold wanted to induce a golden coloration. For this he used salpetre, sulphur, lead and oil. Then he put these ingredients in a container made of copper, which he sealed completely, exactly as it should be done, but when he put it on fire and the container became hot, it burst into many pieces.[19]

The *Feuerwerckbuch* witnesses material production based on practical chemistry in terms borrowed from alchemical transmutation, or more precisely, the making of gold. The analogy was not always with chrysopoeia as such. Vasari's already mentioned fashioning of Van Eyck as an alchemist shows that the image of alchemy as a field of expert inquiry on materials and material transformation was considered sufficiently positive to identify with. However, we should not forget that Vasari's identification of Van Eyck as an alchemist served the purpose of downplaying Netherlandish art as *techne*, only a first step in a narrative of art historical progress culminating in Vasari's beloved Florentine art.

The epistemic value of *techne* was shifting at the time of Vasari. Artisans came to be considered experts of nature likening the artisanal processes of material transformation undertaken in their workshops to those of nature.[20] As Andrea Bernardoni shows in this volume, Biringuccio's *Pirotechnia* is part of this larger process of shifting epistemic value of artisanal processes. Rejecting transmutational alchemy as "false" and the alchemists who practiced it as fraudulent, Biringuccio carved out space for true alchemy as one of the arts of fire. Artisanal workshops, such as Andrea Verrocchio's (c.1435–1488) in which Leonardo da Vinci (1452–1519) apprenticed or Leonardo's own workplace, shared a material culture with alchemical laboratories. As we have already pointed out, artisans used similar equipment and performed "chemical" operations. However, Bernardoni argues, Biringuccio made the claim that these artisans were the true experts on matter, materials and material transformation and that artisanal "chemical" operations were the key to natural knowledge.

One of the readers of Biringuccio's *Pirotechnia* was Johannes Mathesius (1504–1565), a Lutheran preacher in St. Joachimsthal, the center of an important mining district. In her contribution to this volume, Henrike Haug analyzes Mathesius's

---

[19] Hassenstein, *Das Feuerwerckbuch*, 45–6. For translation and discussion, see Werrett, *Fireworks*, 28.

[20] Smith, *Body of the Artisan*. esp. 95–127.

sermons to reveal conceptions of the origin and formation of ores for which Mathesius drew equally on natural philosophy and alchemy and on the artisanal knowledge of the local miners and goldsmiths. Following one of the goldsmiths' most prized objects, a so-called *Handstein*, into the *Kunstkammer*, Haug shows that this knowledge also reached elite collectors. They valued *Handsteine* for the metallogenetic knowledge they embodied, thereby endorsing the shifting epistemic status of the arts.

Epistemic Changes Between Artisans and Alchemists

Biringuccio and Mathesius's St. Joachimsthal are examples of persons and places of exchange between scholarly cultures (in which learning is based on reading and writing) and artisanal cultures (in which learning is based on doing).[21] The laboratories created in Medici Florence, discussed in the chapters by Kieffer and Beretta, were also early examples of such places of exchange, followed by many other European courts. These workplaces facilitated the exchange of materials between alchemy and the arts, and the attraction of the courts also made artisans adopt alchemical language to elevate the status of their craft. However, above all, these court laboratories made possible, well beyond a shared material culture, the exchange of people and knowledge between the arts and alchemy. Although there is a long tradition of experimentation in alchemy and the boundaries between alchemy and art technologies were fluid from the very beginnings of alchemy in Antiquity, in the sixteenth and seventeenth centuries artisans became more deeply involved in alchemical pursuits, and some crafts relied on chemical expertise offered by scholars trained as alchemists.[22] Above all, texts and books, products and symbols of scholarly culture, played an increasingly important role in laboratories and workshops. In these workplaces, a sort of hybrid figure was at work, with one foot in artisanal culture and another in scholarly culture and impossible to categorize in mutually exclusive categories of the scholar and the craftsman.[23] Certain types of crafts—glassmaking, gold- and silversmithing, and porcelain production—seem to have been particularly prone to exchanges between artisanal and scholarly alchemical cultures. By the seventeenth century, the expertise of some glassmakers, silver- and goldsmiths and producers of porcelain was just as based in the worlds of alchemical and bookish learning as it was grounded in hands-on work in the laboratory.[24] Lawrence Principe and Morgan Wesley discuss two examples of such arts: silversmithing and porcelain production, respectively.

---

[21] For 'trading zones' between artisanal and scholarly cultures, see Long, *Artisan/Practitioners*.

[22] Klein, "Chemical Experts."

[23] On hybrid experts, see Klein, "Artisanal-Scientific Experts," and "Depersonalizing the Arcanum."

[24] For glassmaking (not discussed here), see Beretta, *Alchemy of Glass*; and Dupré, "Value of Glass."

Silver- and goldsmiths seem obvious candidates when we think of careers that cross artisan and alchemist. Silversmiths, goldsmiths and alchemists worked on the same materials (silver and gold), and silver- and goldsmiths' expertise was called for to assay the purported gold and silver transmuted by alchemists. Silver- and goldsmiths could also become involved themselves in transmutational endeavours. In his contribution to this book, Principe focuses on the brothers Anthoni and Andries Grill (1604–1655) as examples of such hybrid figures. They ran successful silversmithing businesses in mid-seventeenth century Amsterdam and The Hague. Anthoni Grill's Amsterdam laboratory also produced work on transmutation for which he gained some notoriety. Moreover, Grill's work on chymistry was not only based on learning by doing; he is reported to have read Paracelsus and Glauber. Grill moved between artisanal and scholarly worlds and between learning by doing and by reading. Both came together in his laboratory in Amsterdam.

In the final chapter of this book, Wesley investigates the production of porcelain in the seventeenth century. The potter's art remained outside the canon of the decorative arts until the emergence of the trade of luxury pottery ware in the fifteenth century and Biringuccio's inclusion of pottery among the arts of fire in the sixteenth century. Although (as already mentioned) the Medici sought to imitate the porcelain imported from China in the *Casino*, European hard-paste porcelain, almost identical to that of China, was due to the efforts of seventeenth-century chymists who were both at home in the artisanal world of pottery and in the learned world, according to Wesley. His chapter highlights John Dwight (1633–1704), trained in Robert Boyle's (1627–1691) laboratory and apt with excellent language skills, and Ehrenfried Walther von Tschirnhaus (1651–1708), mathematician and director of the Dresden court laboratory of the Saxon Elector Friedrich August I (1670–1733).[25] Tschirnhaus collaborated with the alchemist and apothecary Johann Friedrich Böttger (1682–1719), also involved in transmutational endeavors at the Dresden court, towards the production of porcelain. In the first porcelain manufactory in Meissen, established by the Elector in his castle Albrechtsburg in 1710, Böttger was the expert overseeing the manufacturing process, knowledgeable in chemistry and skilled in porcelain production. Remarkably, the success of porcelain production was due to the introduction of techniques of investigation, which did not belong to the traditional potter's skills, but to chymistry, which had overlapped with the worlds of metallurgy and assaying for centuries.

Grill, Dwight, and Tschirnhaus are only a few of the numerous examples of glassmakers, silver- and goldsmiths and producers of porcelain whose expertise was both based in the worlds of bookish learning and grounded in hands-on work. Difficult to categorize as either craftsman or scholar, they are responsible for epistemic exchanges between the artisanal and the scholarly worlds. The rise of these hybrid artisan-scholars was connected to the establishment of laboratories in which art technologies and alchemy were brought together, the first of these being at Renaissance courts. Laboratories and artisanal workshops shared a material

[25] See also Pietsch, "Tschirnhaus," and Klein, "Chemical Experts."

culture. In fact, in the early modern period, they were often so similar in terms of the instruments they contained that they carried the same name. This material culture remained largely the same. A change took place at another level: the increasing presence of books and texts in laboratories and workshops populated by figures who merged artisanal and scholarly cultures. This book suggests that this shift in workshop culture facilitated the epistemic exchanges between alchemists and producers of the decorative arts.

Acknowledgements

This book was conceived and written in dialogue with an exhibition project "Art and Alchemy. The Mystery of Transformation." The exhibition is on view at the Museum Kunstpalast in Düsseldorf from April to August 2014. I would like to thank my co-curators of the show, Dedo von Kerssenbrock-Krosigk (director of the Glasmuseum Hentrich in Düsseldorf and the initiator of the exhibition project) and Lawrence Principe, for their inspiration, sound advice, and help with almost every single step in the process of making this book. Without them, the book would have been very different, and more importantly, far below the quality of the present result. The exhibition project in Düsseldorf was a constant inspiration, many thanks to the exhibition team and the curator of the modern part of the show, general director of the Museum Kunstpalast Beat Wismer, who all contributed to making the project such an intellectually rewarding experience.

We worked towards this book during several smaller meetings and one larger conference hosted at the Max Planck Institute for the History of Science in March 2013. Although this book contains only a selection of the papers presented at this conference, I would like to thank all presenters as well as all commentators and discussants for their contributions before during and after the conference. Their interventions have made this a much better book. At the risk of forgetting someone, I would like to name and thank Marjolijn Bol, Suzanne Butters, David Brafman, Spike Bucklow, Mark Clarke, Teresa Esposito, Christine Goettler, Matthew Hunter, Stephen Johnston, Didier Kahn, Martin Kemp, Ursula Klein, Karin Leonhard, Wolfgang Lefèvre, Pamela Long, Alexander Marr, Tine Meganck, Bruce Moran, William Newman, Tara Nummedal, Doris Oltrogge, Valentina Pugliano, Jennifer Rampling, Marlise Rijks, Maurice Yves-Christian Sass, Lorenz Seelig, Pamela Smith, Anke Timmermann, Barbara Tramelli, Berit Wagner, Arie Wallert, Ittai Weinryb, Steve Wharton, and Alan Williams.

This book would not have been possible – and even less possible in such a short time – without the efficient and magnificent editorial assistance of Gina Partridge Grzimek, Lea Poeschl and Celine Camps. My final thanks go to Jed Buchwald for his trust in the project and for considering the book for his series Archimedes, and to Lucy Fleet for her assistance, advice and patience throughout the several stages of this edited book project.

Berlin, Germany Sven Dupré

Bibliography

Beretta, Marco. 2009. *The Alchemy of Glass. Counterfeit, Imitation and Transmutation in Ancient Glass-Making*. Sagamore Beach: Science History Publications.

Davis, Lucy. 2009. Renaissance Inventions: Van Eyck's Workshop as a Site of Discovery and Transformation in Jan van der Straet's Nova Reperta. In *Envisioning the Artist in the Early Modern Netherlands. Netherlands Yearbook for History of Art* 59, eds. H. Perry Chapman and Joanna Woodall, 223–248. Zwolle: Waanders Publishers.

Dupré, Sven. Forthcoming. The Value of Glass and the Translation of Artisanal Knowledge in Early Modern Antwerp. In *Trading Values, Netherlands Yearbook for Art History*, eds. Bart Ramakers, Christine Göttler, and Joanna Woodall.

Evans, R.J.W. 1984. *Rudolf II and his World: A Study in Intellectual History, 1576–1612*. Oxford: Oxford University Press.

Ganzenmüller, W. 1956. Das chemische Laboratorium der Universität Marburg im Jahre 1615. In *Beiträge zur Geschichte der Technologie und der Alchemie*, ed. W. Ganzemüller, 314–322. Weinheim: Verlag Chemie.

Göttler, Christine. 2013. The Alchemist, the Painter, and the "Indian Bird": Joining Arts and Cultures in Seventeenth-Century Antwerp. Adriaen van Utrecht's Allegory of Fire in the Royal Museums of Fine Arts in Brussels. In *Synergies in Visual Culture. Bildkulturen im Dialog. Festschrift für Gerhard Wolf*, eds. Manuela De Giorgi, Annette Hoffmann, and Nicola Suthor, 499–512. Munich: Wilhelm Fink.

Hannaway, Owen. 1986. Laboratory Design and the Aim of Science: Andreas Libavius versus Tycho Brahe. *Isis* 77: 585–610.

Hassenstein, Wilhelm. 1943. *Das Feuerwerckbuch von 1420: 600 Jahre deutsche Pulverwaffen und Buchsenmeister*. Munich: Deutschen Technik.

Howard, Rio. 1983. *La Bibliothèque et le laboratoire de Guy de La Brosse au Jardin des Plantes à Paris*. Genève: Droz.

Joly, Bernard. 1992. Qu'est-ce qu'un laboratoire alchimique?. *Cahiers d'histoire et de philosophie des sciences* 40: 86–102.

Klein, Ursula. 2007. Apothecary Shops, Laboratories and Chemical Manufacture in Eighteenth-Century Germany. In *The Mindful Hand: Inquiry and Invention from the Late Renaissance to Early Industrialisation*, eds. Lissa Roberts, Simon Schaffer, and Peter Dear, 247–276. Amsterdam: Koninklijke Nederlandse Akademie van Wetenschappen.

Klein, Ursula. 2008. The Laboratory Challenge: Some Revisions of the Standard View of Early Modern Experimentation. *Isis* 99: 769–782.

Klein, Ursula. 2012. Artisanal-Scientific Experts in Eighteenth-Century France and Germany. *Annals of Science* 69: 303–306.

Klein, Ursula. 2013. Chemical Experts at the Royal Prussian Porcelain Manufactory. *Ambix* 60: 99–121.

Klein, Ursula. Forthcoming. Depersonalizing the Arcanum. *Technology and Culture*.

Libavius, Andreas. 1606. Commentariorum alchymiae Andreae Libavii Med. D. Pars prima, sex libris declarata. In *Alchymia. Andreae Libavii recognita, emendata, et aucta*. Frankfurt: J Saurius.

Long, Pamela O. 2011. *Artisan/Practitioners and the Rise of the New Sciences, 1400–1600*. Corvallis: Oregon State University Press.

Mander, K. van. 1604. *Het schilder-boeck waerin voor eerst de leerlustighe ieught den grondt der edel vry schilderconst in verscheyden deelen wort voorghedraghen*. Haarlem: Paschier van Wesbusch.

Martelli, Matteo. 2011. Greek Alchemists at Work: "Alchemical Laboratory" in the Greco-Roman Egypt. *Nuncius* 26: 271–311.

Melion, W.S. 1991. *Shaping the Netherlandish Canon. Karel van Mander's Schilder-boeck*. Chicago/London: University of Chicago Press.

Moran, Bruce T. 1991. *The Alchemical World of the German Court: Occult Philosophy and Chemical Medicine in the Circle of Moritz of Hessen (1572–1632)*. Stuttgart: Franz Steiner Verlag.

Newman, William R. 1999. Alchemical Concealment and Symbolism: The Chemical House of Libavius. In *The Architecture of Science*, eds. Peter Galison and Emily Thompson, 59–77. Cambridge, MA: MIT Press.

Newman, William R. 2004. *Promethean Ambitions: Alchemy and the Quest to Perfect Nature*. Chicago: University of Chicago Press.

Newman, William R., and Lawrence M. Principe. 1998. Alchemy vs. Chemistry. The Etymological Origins of a Historiographic Mistake. *Early Science and Medicine* 3: 32–65.

Nummedal, Tara E. 2007. *Alchemy and Authority in the Holy Roman Empire*. Chicago: University of Chicago Press.

Nummedal, Tara E. 2011. Words and Works in the History of Alchemy. *Isis* 102: 330–337.

Osten, Sigrid von. 1998. *Das Alchemistenlaboratorium Oberstockstall. Ein Fundkomplex des 16. Jahrhunderts aus Niederösterreich*. Innsbruck: Universitätsverlag Wagner.

Pereira, Michela. 1997. Utili segreti. Strumenti per lambiccare e acque distillate nei Discorsi. In *Pietro Andrea Mattioli, Siena 1501–Trento 1578. La vita. Le opere. Con l'identificazione delle piante*, ed. Sara Ferri, 281–301. Perugia: Quattroemme.

Pietsch, Ulrich. 2001. Tschirnhaus und das europäische Porzellan. In *Ehrenfried Walther von Tschirnhaus (1651–1708): Experimente mit dem Sonnenfeuer*, eds. P. Plassmeyer and S. Siebel, 68–74. Dresden: Staatliche Kunstsammlungen.

Principe, Lawrence M. 2013. *The Secrets of Alchemy*. Chicago: University of Chicago Press.

Principe, Lawrence M., and Lloyd DeWitt. 2002. *Transmutations: Alchemy in Art*. Philadelphia: Chemical Heritage Foundation.

Rouaze, Isabelle. 1989. Un atelier de distillation du moyen âge. *Bulletin archéologique du Comité des Travaux Historiques et Scientifiques* 22: 159–271.

Shackelford, Jole. 1993. Tycho Brahe, Laboratory Design, and the Aim of Science: Reading Plans in Context. *Isis* 84: 211–230.

Shapin, Steven. 1998. The House of Experiment in Seventeenth-Century England. *Isis* 79: 373–403.

Smith, Pamela H. 1994. *The Business of Alchemy: Science and Culture in the Holy Roman Empire*. Princeton: Princeton University Press.

Smith, Pamela H. 2004. *The Body of the Artisan: Art and Experience in the Scientific Revolution*. Chicago: University of Chicago Press.

Smith, Pamela H. 2006. Laboratories. In *The Cambridge History of Science, 7 vols, vol. 3, Early Modern Science*, eds. Lorraine Daston and Katharine Park, 289–305. Cambridge: Cambridge University Press.

Soukup, Rudolf Werner, and Helmut Mayer. 1997. *Alchemistisches Gold: Paracelsistische Pharmaka: Laboratoriumstechnik im 16. Jahrhundert. Chemiegeschichtliche und archäometrische Untersuchungen am Inventar des Laboratoriums von Oberstockstall/Kirchberg am Wagram*. Vienna: Bohlau Verlag.

Soukop, Rudolf Werner, Sigrid von Osten, and Helmut Mayer. 1993. Alembics, Cucurbits, Phials, Crucibles: A 16th-Century Docimastic Laboratory Excavated in Austria. *Ambix* 40: 25.

Werrett, Simon. 2010. *Fireworks: Pyrotechnic Arts and Sciences in European History*. Chicago/London: University of Chicago Press.

Weyer, Jost. 1992. *Graf Wolfgang II. von Hohenlohe und die Alchemie: alchemistische Studien in Schloß Weikersheim; 1587–1610*. Sigmaringen: Thorbecke.

About the Authors

Marco Beretta is Professor of History of Science at the University of Bologna and editor of *Nuncius – Journal of the Material and Visual History of Science*. He has published extensively on the history of early modern chemistry. His book *The Alchemy of Glass* (2009) was awarded the 2013 Paul-Bunge Prize.

Andrea Bernardoni is a researcher at the Institute and Museum of History of Science (Museo Galileo) in Florence and member of the research group *Artes Mechanicae*, which coordinates numerical simulation and physical reconstruction of historical technological processes in metallurgy and mechanics. His recent publications are: *La conoscenza del fare* (2011); *Il codice Arundel di Leonardo: ricerche e prospettive* (2011); *Esplosioni, fusioni e trasmutazioni, Il monumento a Francesco Sforza e le arti chimiche nel Codice Atlantico di Leonardo da Vinci* (2013).

Sven Dupré is Professor of History of Knowledge at the Institute for Art History at the Freie Universität Berlin and Research Group Director at the Max Planck Institute for the History of Science, Berlin. He has published on a wide range of topics in the history of early modern science, technology and art in Italy, the Spanish Netherlands, the German lands, Britain and France. Recent publications include *Translating Knowledge in the Early Modern Low Countries* (LIT, 2012), *From Earth-Bound to Satellite* (Brill, 2012) and *Silent Messengers* (LIT, 2011).

Henrike Haug is an art historian and works as a research and teaching assistant at the Technical University, Berlin. Her dissertation focussed on forms and media of historical memory in medieval sea republics (Pisa, Genoa, Venice). Her current research interest centres around early modern goldsmiths and their role within the scholarly network of early modern natural philosophers. Her recent publications include: "Materie als Prinzip und Ursache der Individuation. Ähnlichkeit und

Bildnis in der Plastik des 13. Jahrhunderts" (2012); "Wunderbarliche Gewechse. Bergbau und Goldschmiedekunst im 16. Jahrhundert" (2012).

Fanny Kieffer is an associate researcher at the Centre d'Études Supérieures de la Renaissance in Tours and teaches art history at the universities of Tours and Poitiers. She has been a Vittore Branca fellow of the Fondazione Cini and a Frances Yates fellow at the Warburg Institute. Her primary research focus is the interaction between arts and sciences in Renaissance Italy and France. She is about to publish a book on Ferdinando de Medici's patronage and the Uffizi (Brepols, 2014), and has published many articles on the history of medicine and art in late Renaissance Florence.

Matteo Martelli (PhD in Classics, 2006, University of Bologna; PhD in History of Science, 2012, University of Pisa) is currently working at the Humboldt Universität zu Berlin as research associate within the programs "Medicine of the Mind, Philosophy of the Body" and "Der Transfer medizinischer Episteme in den enzyklopädischen Sammelwerken der Spätantike," both under the supervision of Professor Philip van der Eijk. The main topics of his research are history of Greek, Syriac and Arabic alchemy, Galen's pharmacology, and Byzantine medicine (with particular focus on Aetios of Amida).

Sylvie Neven earned her PhD in History, Art and Archaeology at the University of Liège (January 2011). She is currently a postdoctoral fellow of the Fonds National de la Recherche Scientifique (FRS-F.N.R.S.) and lecturer at the University of Liège, Belgium. Since October 2011, she is also a Visiting Scholar at the Max Planck Institute for the History of Science (Berlin), where she collaborates with the Max Planck Research Group "Art and Knowledge in Pre-Modern Europe" (Director, Sven Dupré). Sylvie's work combines historical, philological, experimental and archaeometrical approaches to focus on written sources in art technology.

Lawrence M. Principe is the Drew Professor of Humanities at Johns Hopkins University in the Department of History of Science and Technology and the Department of Chemistry. He is the author of *The Aspiring Adept: Robert Boyle and His Alchemical Quest* (1998), *The Scientific Revolution: A Very Short Introduction* (2011), and *The Secrets of Alchemy* (2013). In 2005, he received the Francis Bacon Medal for his contributions to the history of science, and is currently completing a study of the transformations of chymistry at the Académie Royale des Sciences in the period 1666–1730.

Morgan Wesley is a Doctoral student in History at the University of Oxford. His thesis, *Changing Trends in Ceramic Innovation in Medieval and Early Modern Europe* looks to relocate the examination of ceramic development from an art historical or archaeological pursuit into consilience with the history of science. He has previously lectured on ceramic history for the Sotheby's Institute of Art – London and contributed to the publication of *Art and Authenticity* at this time. His work for *Laboratories of Art* originated from research done at the Huntington Library as a Leeds Hoban Linacre-Huntington Exchange Fellow in 2011.

The Alchemical Art of Dyeing: The Fourfold Division of Alchemy and the Enochian Tradition

Matteo Martelli

Abstract What is Graeco-Egyptian alchemy? Which kinds of techniques and craft practices does it encompass? And what were its goals? The paper addresses these questions by investigating the earliest Greek alchemical texts preserved both in Byzantine and in Syriac manuscripts. Already during the first centuries AD, in the Graeco-Roman Egypt it is possible to recognize some disagreement over the definition of alchemy and its expected outcomes. On the one hand, ps.-Democritus's four books and the Leiden and Stockholm papyri support a fourfold division of alchemy including processes for making gold, silver, and precious stones (glass working included), and for dyeing wool purple. On the other hand, Isis's treatise focuses only on the making of precious metals, which is identified with the main goal of alchemy during the late Byzantine tradition. In the process that led to such a simplification of the technical background of alchemy Zosimus's work seems to represent an important turning point. In fact the author inherited the above mentioned polarity and discussed different ideas of alchemy in a key text (here edited and translated into English for the first time) on the revelation of alchemy based on the Enochian myth of the fallen angels.

Alchemy and the Making of Gold: An Overview of the Byzantine Tradition

Between the seventh and the eleventh century different people—including imperators, caliphs, and scholars moved by antiquarian curiosities—became interested in *chēmeia* (alchemy) and somehow identified this art with *chrysopoeia* (the making of gold) and *argyropoeia* (the making of silver). Sure enough, these people were not practitioners and they did not test the techniques described by alchemical recipes.

M. Martelli (✉)
Department of Classics, Humboldt Universität zu Berlin, Berlin, Germany
e-mail: martellm@cms.hu-berlin.de

S. Dupré (ed.), *Laboratories of Art*, Archimedes 37,
DOI 10.1007/978-3-319-05065-2_1,

However, if we consider that the Byzantine anthologies preserving most of the Greek and Byzantine alchemical writings were probably compiled from the seventh century onwards, we should wonder to what extent a similar attitude could have affected the selection of the works transmitted by these collections.[1] Did such a focus on the making of precious metals influence the choice of the treatises to be included into Byzantine anthologies and translated into Syriac and Arabic? Although a complete answer to this question would surpass the aim of this chapter and would require a wider analysis of the hundreds of alchemical manuscripts nowadays kept in several European and Middle-Eastern libraries, in the following I focus my attention especially on the polarity between the idea of alchemy as *chrysopoeia* attested by Byzantine scholars and a broader interest in different dyeing techniques attested by the most ancient alchemical writings (first to fourth century AD). Some of these works, in fact, covered a wider set of techniques focusing not only on how to dye metals yellow and white (that is, how to transform them into gold and silver), but also on how to dye crystal quartz different colours in order to produce precious stones, and how to dye wool purple.

In mid-eleventh century Constantinople, the Byzantine scholar Michael Psellos (1018–c.1078) composed a small treatise on the making of gold in the form of a letter addressed to the Patriarch Michael I Cerularius (1043–1059), who asked the philosopher to make an investigation on the ancient alchemical methods of χρυσοποιία (*chrysopoeia*). After some theoretical remarks, Psellos introduced and somehow justified the specific focus of his small treatise as follows (*Letter on the Making of Gold*, § 5 *partim*):

> Since in my preface I have already insisted enough on the fact that transformations of matter happen according to natural changes, and not by means of magic spells, miracles, or some other secret practice (so, we must not wonder), it is time to pass on to this art of transformation. I would have liked to compose a complete discourse on this art and on how to work the matters [...] and to teach what makes quartz and sapphire porous, what produces a fake emerald and beryl, which nature can soften stones, which one can dilute pearls and make them watery, and which one can make then again solid and round, and how to whiten them [...] However, since you [Michael Cerularius] do not allow me to delay with such superfluous inquiries, wasting all my studiousness in a worthless research, but you want me to examine with which substances and according to which scientific method gold may be produced, I am going to explain only this topic.[2]

[1] See Mertens, "Greco-Egyptian Alchemy," 220–5.

[2] Edition by Bidez, *Épitre sur la Chrysopée*, 30, 16–31, 9: "Ἐπεὶ οὖν ἱκανῶς ἡμῖν πεπροοιμίασται ὡς αἱ τῶν ὑλῶν μεταβολαὶ φυσικήν τινα ἀλλοίωσιν ἔχουσιν καὶ οὐκ ἐξ ἐπῳδῆς τινος ἢ τερατείας ἢ ἄλλης ἀρρητουργίας (διὸ καὶ θαυμάζειν οὐ χρή), ἐπ' αὐτὴν ἤδη σοι τὴν τέχνην χωρῶ τῆς μεταβολῆς. Ἐβουλόμην μὲν οὖν καθολικήν τινά σοι τεχνολογίαν ποιήσασθαι καὶ πᾶσαν ὑλουργίαν διερευνήσασθαι, [...] διδάξαι τε τί μὲν τὸ τὸν κρύσταλλον ἀραιοῦν, τί δὲ τὸ τὸν ὑάκινθον, καὶ πῶς ἄν τις καὶ σμάραγδον οὐκ ὄντα ποιήσῃ καὶ βήρυλλον, τίς δὲ ἡ φύσις τοῦ τὰς λίθους ἁπάσας μαλάττοντος, καὶ πῶς μὲν ἡ μαργαρῖτις λυθείη καὶ εἰς ὕδωρ ἀναλυθείη, πῶς δ' αὖθις συμπαγείη καὶ σφαιρωθείη, τίς δὲ ὁ λόγος τῆς τούτων λευκάνσεως [...]· ἐπεὶ δὲ σὺ σχολάζειν ἡμᾶς ἐν τοῖς περιττοῖς οὐκ ἐᾷς οὐδὲ ἐν τοῖς ἀσπουδάστοις καταναλίσκειν πᾶν τὸ φιλότιμον, τοῦτο δὲ μόνον διερευνῆσαι προῄρησαι ἐκ τίνων ὑλῶν καὶ διὰ ποίας τῆς ἐπιστήμης χρυσὸν ἄν τις ποιήσειε, ταύτην μόνην τὴν τεχνολογίαν σοι δίειμι." For an Italian commented translation, see Albini, *Michele Psello*.

The passage comes after a long introduction (§§ 1–4) in which Psellos rhetorically discusses the philosophical framework within which *chrysopoeia* is to be understood and assessed. At the very beginning (§ 1, ll. 5f.) the scholar identifies the main topic of his letter with ἡ ἐμπύριος τέχνη (the art of fire), whose first aim is said to be the transformation of lead, tin, and any other metal into gold (§ 1, ll. 12–15). The key concept of transformation is framed and expanded in the following paragraphs (§§ 2–4), where Psellos tries to set the basic principles guiding his κατ'ἐπιστήμης (scientific) explanation of this art. He insists on the four elements that compose the physical world, on their reciprocal changes, and on the alterations of the four qualities (namely wet, dry, hot, and cold). These changes are explained through examples taken from various meteorological phenomena mainly based on evaporation and condensation of water.[3] For instance, "water becomes like a stone [i.e. changes into earth] when it freezes [and] becomes air when it evaporates."[4] Moreover, similar transformations of the basic elements/qualities also help to understand more complex phenomena, such as the case of a tree burned by a lightning strike, in which all the humidity of wood is consumed by the fire.[5] In Psellos's opinion, φυσικὴ ἀλλοίωσις (natural processes of alteration) of the same kind are to be detected behind the "art of fire," which is redefined as ἡ τέχνη τῆς μεταβολῆς (the art of transformation) at the beginning of the section quoted above (ll. 3–4). This art, in fact, included a wide set of practices and techniques dealing with the transformation of various ὕλαι (materials), such as metals, stones, and glasses. However, while a similar approach would have driven Psellos to consider all the methods listed in the passage, the narrower interests of Cerularius forced the scholar to focus his inquiry only on the making of gold.

In this respect, the attitude of Psellos's patron seems to reflect a more general idea about alchemy that is confirmed by other Byzantine sources. First of all, various encyclopaedia and lexica provide a similar definition of the difficult term *chēmeia* (χημεία or χυμεία), probably introduced for the first time by Zosimus (third century AD) in reference to the art of dyeing metals.[6] In the tenth century, the lexicon *Suda* (χ 280 Adler) explained the term *chēmeia* as ἡ τοῦ ἀργύρου καὶ χρυσοῦ κατασκευή (the preparation of gold and silver).[7] Outside the Byzantine

[3] Meteorology represents an important aspect of Psellos's natural philosophy; the scholar devoted to this topic both many chapters of his *De Omnifaria Doctrina* (§§ 139–78 in Westerink, *Omnifaria Doctrina*) and some specific essays (see Bidez, *Épitre sur la Chrysopée*, 51–70; and Duffy, *Michaelis Pselli*, texts 19–31). On the relations between Psellos's alchemical interests and his investigation of the physical world, see Katsiampoura, "Transmutation of Matter," 665–7.

[4] "τὸ μὲν γὰρ ὕδωρ πηγνύμενον λιθοῦται εἰς κρύσταλλον [and] τοῦτο εἰς ἀτμίδα λυόμενον ἀὴρ καθίσταται."

[5] In § 4 Psellos explains the petrifaction of an oak struck by a lightning strike: a quick and sharp lightning does not only make the oak black, but consumes all its humidity and transforms the wood into stone.

[6] See *infra*, § 3.

[7] The entry continues by telling how the emperor Diocletian (284–305) made to burn all the Egyptian books on alchemy (in the Greek text: "περὶ χημείας χρυσοῦ και ἀργύρου"); this information probably depends on the *Chronicle* (*Ἱστορία χρονική*) of John of Antioch (active under Heraclius's reign, 610–641): see fragment 280 in Roberto, *Ioannis Antiocheni Fragmenta*.

Empire, the same definition had been already given by the Persian scholar al-Nadīm (ninth century AD), who devoted the tenth chapter of his *Kitāb al-Fihrist* (*Book of Catalogue*) to the *ṣinā'at al-kīmīyā'* (art of alchemy): at the very beginning of the chapter this art is defined again as "the making of gold and silver from other metals."[8] More than one century later, the Syriac scholar Bar Bahlul (eleventh century AD) dealt with the same subject in two entries of his lexicon:

1. Duval, *Lexicon Syriacum*, vol. I, 901:
   Kimiya, the dark stone. It is the work of the art of gold and silver. And there is someone who explains this word from the name *kima* (the constellation of the Pleiades), that is the height stars, because (this art) is accomplished by means of the height mixtures.[9]
2. Duval, *Lexicon Syriacum*, vol. I, 904:
   Kemela'a, that is the alchemy (*kimiya'a*) of the seven bodies and of the various natures; [it includes] arts such as goldsmith's art and copper, iron, and glass working.[10]

Only the last definition includes a wide set of techniques applied to different substances (among which glass is mentioned), while most of the above-quoted texts identifies alchemy with the making of gold and silver. The two precious metals were probably the most attractive outcomes that anyone engaged in alchemical practices tried to achieve out of this art, as is possible to infer from a few accounts on alchemists preserved by Byzantine and Syriac chronographers. In the mid-sixth century, for instance, John Malalas (491–578) reported the story of John Istmeos, χειμευτὴς ὑπάρχων καὶ φοβερὸς ἐπιθέτης (alchemist and tremendous impostor), who moved from Antioch to Constantinople, where he was condemned by the emperor Anastasios (491–518) because of the adulterated gold objects he tried to

[8] I have quoted the translation proposed by Dodge (*The Fihrist*, 841), who checked several manuscripts (see pp. xxiii–xxxiv) not taken into account by Flügel, who edited the Arabic text. In Flügel's edition (*Kitāb al-Fihrist*, vol. II, 351) the passage reads: زعم اهل ضاعة الكيمياء وهى صنعة الذهب والفضة من غير معادنها, translated by Fück, "Arabic Literature of Alchemy," 88, as follows: "The adepts of the Art of Alchemy, which is the art of making gold and silver without (recourse to) mining."

[9] ܟܝܡܝܐ. ܟܐܦܐ ܟܡܝܪܘܬܐ (*sic*). ܦܘܠܚܢܐ ܕܐܘܡܢܘܬܐ ܕܕܗܒܐ ܘܣܐܡܐ. ܘܐܝܬ ܕܡܦܫܩ ܠܗ ܡܢ ܫܡ ܟܝܡܐ. ܗ̄ ܟܘܟܒܐ ܬܡܢܝܐ ܗ̄ ܡܬܓܡܪܐ ܡܢ ܬܡܢܝܐ ܡܘܙܓܐ. The expression ܟܐܦܐ ܟܡܝܪܘܬܐ literally means 'stone of sadness.' Berthelot & Duval, *Chimie*, 133 translated it as 'pierre philosophale.' Instead of ܟܡܝܪܘܬܐ ('sadness, grieving'), I read ܟܡܝܪܬܐ ('black, dark'); on the tendency of identifying 'alchemy' with the substance used for transforming vile metals into gold and silver, see Gildemeister, "Alchymie."

[10] ܩܡܠܝܐ ܐܘܟܝܬ ܟܝܡܝܐ ܗ̄ الكيميا ܕܫܒܥܐ ܓܘܫܡܝܢ ܘܕܟܝܢ̈ܐ ܡܫܚܠܦܐ. ܐܘܡܢ̈ܘܬܐ ܐܝܟ الصناعات مثل الصياغة وعمل الشبة و الحديد و الزجاج. The term ܩܡܠܝܐ is often used in the lexicon (Duval, *Lexicon Syriacum*, vol. I, 78,12; 114,14; 222,5 and 20; 239,5; and vol. III, 160, *s.v.* ܦܠܘܣܐ), especially in the expression ܦܠܘܣ ܩܡܠܝܐ. In Duval's opinion, the term is an early corruption for ܟܝܡܝܐ.

sell as authentic (*Chronographia* XVI 5). The Syrian city of Antioch was again the scene of an 'alchemical affaire' at the beginning of the eighth century, when Isaac, bishop of Ḥarrān, is said to have become archbishop of the city (755–756) thanks to his alchemical knowledge. According to various Syriac chronicles, Isaac murdered a monk of Edessa and stole from him a powder that, when applied on lead, could transform it into gold.[11] By means of this powder, Isaac was able to find favour with the Caliph al-Mānṣur (754–775), who turned out to be very interested in this substance and supported (at least for a little while) Isaac's political career. Moreover, Arabic sources claim that the same caliph sent his emissary 'Umāra ibn Ḥamza to the court of Constantine V (741–775), who led the guest into a big storehouse containing two kinds of powder: a white elixir that could transmute metals into silver, and a yellow/red one that could transform metals into gold.[12]

Two Early Alchemical Treatises: Ps.-Democritus and Isis

After the above-discussed passage, Psellos carries on his investigation by listing and commenting seven recipes describing how to prepare specific φάρμακα (medicines) with which to treat base metals (lead first of all) and transform them into gold. The section is concluded by the rhetorical question: "How so? Should I shortly reveal you all the wisdom of Democritus without leaving anything in the innermost sanctuary?"[13]

The reference to the Ἀβδηριτικὴ σοφία (the wisdom of Democritus), the philosopher from Abdera *par excellence*, is noteworthy, since Democritus (fl. 440–380 BC) is usually considered one of the most ancient authors of alchemical treatises handed down by Byzantine manuscripts. These treatises are obviously spurious and were composed by an anonymous author in the second half of the first century AD. However, in contrast with Psellos's account focusing only on *chrysopoeia*, the alchemical tradition ascribes to the philosopher four books dealing with a wider set of interests that included not only the making of silver and gold, but also the making of precious stones and the purple dyeing of wool. For instance, the alchemist Synesius (fourth century AD), who wrote a commentary on ps.-Democritus's

[11] See (a) the fourth part of the *Zuqnīq Chronicle* or *Chronicle of ps.-Dyonisus of Tell-Mahre*, edited by Chabot, *Pseudo-Dyonysianum*, 66; (b) the chronicle of Michael the Syrian, edited by Chabot, *Michel le Syrien*, 474; (c) the anonymous chronicle edited by Brooks, "Syriac Fragment," 217–8.

[12] For instance, the story is told by the geographer Ibn al-Faqīh al-Hamaḏani (Goeje, *Ibn al-Faqīn al-Hamaḏānī*, 137–9; French transl. by Massé, *Abrégé du Livre de Pays*, 164–6); see Strohmaier, "Umāra ibn Ḥamza," 21–2.

[13] Bidez, *Épitre sur la Chrysopée*, 40,7–8: "Τί οὖν; πᾶσάν σοι τὴν Ἀβδηριτικὴν σοφίαν ἀνακαλύψομεν ἐν βραχεῖ καὶ οὐδὲν ἐντὸς τοῦ ἀδύτου ἀφήσομεν."

alchemical writings, introduced the alleged production of the philosopher as follows (Syn. Alch. § 1, ll. 5–14 Martelli):

> So now let me start telling you something about this man, the philosopher Democritus, who was a natural philosopher from Adbera, who investigated every natural question and explained every phenomenon on the basis of its own nature. Abdera is a Thracian town; but he became a very wise man when he came to Egypt and was initiated by the great magician Ostanes with all Egyptian priests. He got from Ostanes the bases [of the alchemical processes] and wrote four books on dyeing, that is, on gold and silver and precious stones and purple.[14]

The great relevance ps.-Democritus's four books had in the later alchemical tradition—that is unanimous in recognizing the atomist as one of the founders of this discipline—allows us to consider the four areas of expertise included in his work as relevant parts of the technical background from which alchemy took its first steps in the first-second century AD. The four books are concisely defined by Synesius as βαφικαὶ βίβλοι (books on dyeing), since ps.-Democritus dealt with a wide range of dyeing techniques based on various methods for changing the colours of metals, producing precious stones and using purple substitutes. The same areas of expertise are covered by the Graeco-Egyptian papyri of Leiden and Stockholm, which hand down several recipes describing (a) how to dye metals yellow or white (that is, how to transform them into gold or silver), (b) how to dye κρύσταλλος (quartz) and counterfeit different kinds of precious stones out of it, and (c) how to dye wool purple by means of various substitutes of the expensive Tyrian purple.[15]

Moreover, ps.-Democritus combined the technical explanation of these methods with a theoretical approach that emphasized the central role played by dyeing drugs and their properties.[16] He analysed and classified τὰ φυσικά (the natural ingredients) according to their capacities of dyeing and of mixing each other properly.[17] For instance, in the middle of the book on the making of gold, after a long criticism against young people who did not accurately investigate the properties of natural substances, ps.-Democritus claims (*PM* § 16, ll. 178–80 Martelli):

> If these young men had practised for these kinds of investigations, they would not be in trouble, since they could set to work with judgement. But they do not know the antipathies

[14] The Greek text reads: "Ἐν ᾧ οὖν πρόκειται ἡμῖν εἰπεῖν τίς ἄν εἴη ὁ ἀνὴρ ἐκεῖνος, ὁ φιλόσοφος Δημόκριτος, ἐλθὼν ἀπὸ Ἀβδήρων, φυσικὸς ὢν καὶ πάντα τὰ φυσικὰ ἐρευνήσας καὶ συγγραψάμενος τὰ ὄντα κατὰ φύσιν. Ἄβδηρα δέ ἐστι πόλις Θρᾴκης, ἐγένετο δὲ ὁ ἀνὴρ λογιώτατος, ὃς ἐλθὼν ἐν Αἰγύπτῳ ἐμυσταγωγήθη ὑπὸ τοῦ μεγάλου Ὀστάνου ἐν τῷ ἱερῷ τῆς Μέμφεως, σὺν καὶ πᾶσι τοῖς ἱερεῦσιν Αἰγύπτου. Ἐκ τούτου λαβὼν ἀφορμάς, συνεγράψατο βίβλους τέσσαρας βαφικάς, περὶ χρυσοῦ καὶ ἀργύρου καὶ λίθων καὶ πορφύρας." First edition in Berthelot & Ruelle, *Anciens alchimistes grecs*, vol. II, 57 (hereafter *CAAG*).

[15] The most recent edition is by Halleux, *Papyrus de Leyde*; for a general overview on the contents of the two papyri, see, in particular, 13–7. English translation in Caley, "Leiden Papyrus X," and "Stockholm Papyrus."

[16] See, for instance, ps.-Dem. Alch. *PM* § 20, ll. 215–24 in Martelli, *Pseudo-Democrito*, 202–4 (= *CAAG* II 49), 135–48.

[17] On ps.-Democritus's catalogues of dyestuffs, see Martelli, *Pseudo-Democrito*, 83–90.

of natures, how one species upsets ten: a drop of oil has the capacity to remove much purple, and a pinch sulphur may burn many species.[18]

On the other hand, the form in which Byzantine manuscripts have preserved ps.-Democritus's writings does not allow us to reconstruct the exact content of each book and their reciprocal connection. In fact, only an epitomized version has been included in the Byzantine collections, which contain just two sections ascribed to the philosopher:

1. The book on the making of gold (or more probably a summarized version of it) is introduced by a very short section on purple dyeing and handed down under the title of *Φυσικὰ καὶ μυστικά* (*Natural and Secret Questions*).[19]
2. A selection of passages from the book on silver forms the second section entitled *Περὶ τῆς ἀσήμου ποιήσεως* (*On the Making of Silver*).[20]

Some information about the book on stones—otherwise lost in its original Greek version—is provided only by the indirect tradition. In fact, a later recipe book preserved by a few Byzantine manuscripts under the title of *Deep Tincture of Stones, Emeralds, Rubies and Jacinth from the Book Taken out from the Innermost Sanctuary of Temples* includes some passages discussing the dyeing methods applied by ancient alchemists on stones.[21] Along with Democritus, his pretended master Ostanes, Moses, and Maria the Jewish are also quoted.[22] Regrettably, the writings of these ancient authors are lost, so that we cannot understand to what extent they actually dealt with similar topics. However, as far as ps.-Democritus is concerned, in all likelihood he attributed a certain relevance to the dyeing of stones, if it is true that he devoted an entire book to this subject. On the contrary, the exclusion of this book—along with most of the book on purple—from the epitomized version handed down by the Byzantine anthologies evidences the criteria adopted by the epitomiser of ps.-Democritus's writings: he seems to have focused his attention only on the transmutation of metals into gold and silver, so revealing an attitude that tallies with the approach we have already detected in the above-discussed Byzantine sources.

Nevertheless, although the relevance of *chrysopoeia* in Byzantine times can somehow explain the loss of part of ps.-Democritus's books, one of the manuscripts

[18] The Greek text reads (= *CAAG* II 47–8): "Εἰ ἐν τούτοις ὑπῆρχον ἀσκούμενοι οἱ νέοι, οὐκ ἂν ἐδυστύχουν, κρίσει ἐπὶ τὰς πράξεις ὁρμῶντες· οὐ γὰρ ἐπίστανται τὰ τῶν φύσεων ἀντιπαθῆ, ὡς ἓν εἶδος δέκα ἀνατρέπει. Ῥανὶς γὰρ ἐλαίου οἶδε πολλὴν ἀφανίσαι πορφύραν, καὶ ὀλίγον θεῖον εἴδη κατακαῦσαι πολλά."

[19] See Martelli, *Pseudo-Democrito*, 180–205 (and 73–9) = *CAAG* II 41–9.

[20] See Martelli, *Pseudo-Democrito*, 206–16 (and 79–83) = *CAAG* II 49–53.

[21] Edited in *CAAG* II 350–64; the Greek title reads: "καταβαφὴ λίθων καὶ σμαράγδων καὶ ὑακίνθων ἐκ τοῦ ἀδύτου τῶν ἱερῶν ἐκδοθέντος βιβλίου." The earliest testimonies are the manuscripts *Parisinus* gr. 2325 (13th century; fols. 160v–173v), and *Parisinus* gr. 2327 (15th century; fol. 147r–159r).

[22] See, in particular, for Democritus: *CAAG* II 353,11–25, and 354,12–357,19; for Ostanes: *CAAG* II 351,16–28, and 352,10 (fragments reedited by Bidez & Cumont, *Mages hellénisés*, vol. II, 323–4); for Moses: *CAAG* II 353,19; for Maria: *CAAG* II 351,23; 352,2–8; 355,1, and 257,19.

that hands down the above-mentioned recipe book on precious stones preserves another important treatise that introduces a new important element into the discussion. The *Parisinus* gr. 2325 (fols. 256r–258v), in fact, contains a book ascribed to the goddess Isis, which probably dates back to the second century AD and contains an important account of the revelation of the alchemical art. According to the beginning of the treatise, Isis received a revelation from two angels who descended from heaven because they were attracted by the beauty of the goddess (*CAAG* II 28–9):

> *Isis the Prophet to Her Son Horus.* O son, as you were about to leave and fight a battle against the unfaithful Typhon for the kingdom of your father, I went to Hormanouthis [city, sanctuary?] of the holy art in Egypt, where I stayed a long time. According to the recession of the convenient time and to the necessary course of the spherical movement, it happened that one of the angels living in the first firmament, after watching me from above, wanted to have sexual intercourse with me. When he arrived and started taking this direction, I did not give myself, because I wanted to learn the preparation of gold and silver. After I asked him this question, he told me that he was not allowed to reveal this point, because these secrets surpassed him, but (he told me) that the next day his superior, the angel Amnaēl, would come, and that he would be able to give a reply for a similar inquiry [. . .] The next day, his superior Amnaēl appeared when the sun was in the middle of its course, and he came down. Taken by the same desire for me, he did not await, but he hastened to get what he came for; but I was not less focused on what I was searching for. He longed for it, but I did not give myself and I was able to curb his desire until he showed me his mark on the head and revealed all the mysteries I was looking for, without envy and faithfully.[23]

The ἱερὰ τέχνη Αἰγύπτου (Egyptian holy art) quoted at the beginning of the passage is clearly identified by Isis with ἡ τοῦ χρυσοῦ καὶ ἀργύρου κατασκευή

[23] I have followed the more recent edition by Mertens, *Lettre d'Isis*, 128–31: "*Ἶσις προφῆτις τῷ υἱῷ Ὥρῳ*. Ἀπιέναι σου μέλλοντος, ὦ τέκνον, ἐπὶ ἀπίστου Τυφῶνος μάχης καταγωνίσασθαι περὶ τοῦ πατρός σου βασιλείας, γεναμένης μου <πρὸς> Ὁρμανουθί, <. . .> ἱερᾶς τέχνης Αἰγύπτου, καὶ ἐνταῦθα ἱκανὸν χρόνον διέτριβον. Κατὰ δὲ τὴν τῶν καιρῶν παραχώρησιν, καὶ τὴν τῆς σφαιρικῆς κινήσεως ἀναγκαίαν φοράν, συνέβη τινὰ τῶν ἐν τῷ πρώτῳ στερεώματι διατριβόντων, ἕνα τῶν ἀγγέλων, ἄνωθεν ἐπιθεωρήσαντά με, βουληθῆναι τῆς πρὸς ἐμὲ μίξεως κοινωνίαν ποιῆσαι. Φθάσαντος δὲ αὐτοῦ καὶ εἰς τοῦτο γίγνεσθαι μέλλοντος, οὐκ ἐπέτρεπον ἐγώ, πυνθάνεσθαι βουλομένη τὴν τοῦ χρυσοῦ καὶ ἀργύρου κατασκευήν. Ἐμοῦ δὲ τοῦτο αὐτῷ ἐρωτησάσης, <οὐκ> ἔφη ὁ αὐτὸς ἐφίεσθαι περὶ τοῦτο ἐξειπεῖν, διὰ τὴν τῶν μυστηρίων ὑπερβολήν, τῇ δὲ ἑξῆς ἡμέρᾳ παραγίγνεσθαι τὸν τούτου μείζονα ἄγγελον Ἀμναήλ, κἀκεῖνον ἱκανὸν εἶναι περὶ τῆς τούτων ζητήσεως ἐπίλυσιν ποιήσασθαι. [. . .] Τῇ δὲ ἑξῆς ἡμέρᾳ ἐπεμφανίσας καὶ τοῦ ἡλίου μέσον δρόμου ποιοῦντος, κατῆλθεν ὁ τούτου μείζων Ἀμναήλ. Τῷ αὐτῷ περὶ ἐμὲ ληφθεὶς πόθῳ οὐκ ἀνέμενεν, ἀλλ' ἔσπευδεν ἐφ' οὗ καὶ παρῆν· ἐγὼ δὲ οὐχ ἧττον ἐφρόντιζον περὶ τούτων ἐρευνᾶν. Ἐγχρονίζοντος δὲ αὐτοῦ, οὐκ ἐπεδίδουν ἑαυτήν, ἀλλ' ἐπεκράτουν τῆς τούτου ἐπιθυμίας ἄχρις ἂν τὸ σημεῖον τὸ ἐπὶ τῆς κεφαλῆς ἐπιδεικνύηται καὶ τὴν τῶν ζητουμένων μυστηρίων παράδοσιν ἀφθόνως καὶ ἀληθῶς ποιήσηται." The beginning of the story clearly refers to the Egyptian myth of Horus fighting against Seth, the killer of his father Osiris. The identification between the Egyptian god Seth and Typhon is quite common in Hellenistic and Roman Egypt: see, for instance, Plutarch, *De Iside et Osiride* 41 (= 367d): "τὸν Τυφῶνα Σὴθ Αἰγύπτιοι καλοῦσι." – "Egyptians gives to Seth the name of Typhon." On the other hand, the toponym *Ὁρμανουθί* is not clear and not otherwise attested. Probably it refers to an Egyptian city, although it does not match any of the five towns, where alchemy was practiced according to *CAAG* II 26. Various corrections of the form *Ὁρμανουθί* have been proposed: see, in particular, Reitzenstein, *Poimandres*, 141, no. 3 (who proposed *Ὁρμαχουθί*, lit. 'Horus of Edfu'), and Mertens, *Lettre d'Isis*, 56–60 (who proposed *Μενουθί*, that is an area in the Egyptian city of Canopus).

(the preparation of gold and silver), that is, the same definition of alchemy attested by the Lexicon *Suda* (χ 280) and supported by the Byzantine sources that have been briefly analyzed in the first paragraph. The inclusion of this definition in such an ancient text is noteworthy, since it proves that some of the earliest authors already focused their treatises only on metallurgical practices.

Consequently, the last part of Isis's work lists five metallurgical recipes describing: (a) how to make mercury solid and mix it with lead; (b) how to prepare a white dyeing drug by which, according to the third recipe (c), an iron-copper alloy was treated and dyed white (the same alloy is dyed yellow in the second part of the recipe); (d) how to mix the substances prepared according to the first three recipes; (e) how to process a metallic body before dyeing it white.[24] This section is concluded by a more general statement according to which all the οἰκονομίαι (treatments), the δίπλωσις (techniques for doubling the weight of gold or silver objects), and the καταβαφαί (dyeing processes) were moved by the same aim, that was, according to the beginning of the treatise, the making of gold and silver.[25]

Towards a Definition of Alchemy: Zosimus and the Enochian Myth

The plot of Isis's story, although set in a new Egyptian framework, clearly depends on the Enochian account of the fallen angels who taught mankind about a divine and forbidden knowledge that included a wide set of crafts. The myth is fully developed in the very first part of the so-called *Book of Enoch* (or *1Enoch*), one of the *pseudepigrapha* of the Ancient Testament ascribed to Enoch, the grandfather of Noah. The book is composed of different segments (or treatises), and in its most developed form it is preserved only by a translation in Classical Ethiopic (fifth-sixth century AD). However, the major part of it is much earlier, as one can infer from various sections that have been preserved by the Aramaic manuscripts discovered in Qumran's caves.[26] In particular the first part, usually called *The Book of Watchers* (= *1Enoch*, chap. 1–36), is handed down in several Dead Sea scrolls and in all likelihood dates back to the third century BC.[27] Moreover, this book has

[24] Mertens, *Lettre d'Isis*, 134–8 = *CAAG* II 31–3.

[25] Mertens, *Lettre d'Isis*, 138, ll. 113–6 = *CAAG* II 33, § 16.

[26] The secondary literature on *1Enoch* is vast; for a general introduction on its content, see, for example, Knibb, *Book of Enoch*, 7–35. For a recent English translation of the Ethiopic text, see Black, *Book of Enoch*.

[27] These manuscripts have been found in cave 4; the Aramaic text has been edited and translated by Milik, *Books of Enoch, Aramaic fragments*.

been translated into Greek quite early, probably already in the second century BC, within the same cultural context that produced the Septuagint translation of *Daniel*.[28]

At the beginning of *The Book of Watchers* (*1Enoch*, chap. 6–8) the author explains that two hundred angels—called ἐγρήγοροι (watchers) or ἄγγελοι υἱοὶ οὐρανοῦ (the angels, sons of the Heaven) in the Greek version—started desiring the daughters of men and descended from heaven in order to have intercourse with them. The Greek version of the passage, extensively quoted by the chronographer Syncellus, goes on as follows (Sync. p. 12, ll 8–17 Mosshammer):

> The leaders [of these angels] and all the rest [of the two hundred watchers] took for themselves wives in AM 1170, and they began to defile themselves with them up to the Flood. [. . .] And they were increasing in accordance with their greatness, and they taught themselves and their wives the uses of potions and spell. First Azaēl, the tenth of the leaders, taught them to make swords and armours and every instrument of war and how to work the metals of the earth and gold, how to make them into adornments for their wives, and silver. He showed them also the use of cosmetics and beautifying the face and choice stones and colouring tinctures.[29]

The list of crafts revealed by Azaēl, which includes the working of metals (with explicit mention of gold and silver) and of precious stones along with τὰ βαφικά (dyeing procedures) in a more general sense, shows clear similarities with the topics covered by ps.-Democritus's four books, which nevertheless do not refer back to this myth (at least in the preserved sections). On the contrary, the myth was reworked in Isis's treatise, where the secret teaching of the angels was limited only to the preparation of gold and silver. These different approaches to what was considered part of the alchemical art were inherited and discussed by Zosimus, a third century author who clearly reused the Enochian myth in his own account of the origins and developments of alchemy. In fact, Zosimus first introduced the term *chēmeia* with reference to fallen angels' revelation, which was written down in specific and secret books. Regrettably, Zosimus's treatise is lost in its Greek original form, and just its beginning is quoted by Syncellus straight after the above-mentioned passage taken from *1Enoch*. However, a more complete version is preserved by an unedited Syriac translation handed down in the Cambridge manuscript Mm. 6.29. The two versions read as follows:

1. Sync. p. 14, ll. 2–14 Mosshammer:
   But it is also fitting to cite a passage regarding them [i.e. the divine scriptures] from Zosimus, the philosopher of Panopolis, from his writings to Theosebeia in

[28] Baar, "Aramaic-Greek Notes," 191–2.

[29] Translation by Adler & Tuffin, *George Synkellos*, 17. The Greek text reads: "Οὗτοι καὶ οἱ λοιποὶ ἐν τῷ αρο´ ἔτει τοῦ κόσμου ἔλαβον ἑαυτοῖς γυναῖκας καὶ ἤρξατο μιαίνεσθαι ἐν αὐταῖς ἕως τοῦ κατακλυσμοῦ. [. . .] καὶ ἦσαν αὐξανόμενοι κατὰ τῆν μεγαλειότητα αὐτῶν, καὶ ἐδίδαξαν ἑαυτοὺς καὶ τὰς γυναῖκας ἑαυτῶν φαρμακείας καὶ ἐπαοιδίας. πρῶτος Ἀζαὴλ ὁ δέκατος τῶν ἀρχόντων ἐδίδαξε ποιεῖν μαχαίρας καὶ θώρακας καὶ πᾶν σκεῦος πολεμικὸν καὶ τὰ μέταλλα τῆς γῆς καὶ τὸ χρυσίον πῶς ἐργάσωνται, καὶ ποιήσωσιν αὐτὰ κόσμια ταῖς γυναιξί, καὶ τὸν ἄργυρον. ἔδειξε δὲ αὐτοῖς καὶ τὸ στίλβειν καὶ τὸ καλλωπίζειν καὶ τοὺς ἐκλέκτους λίθους καὶ τὰ βαφικά."

the ninth book of *Imouth*, reading as follows:[30] "The Holy Scriptures, that is the books, say, my lady (i.e. Theosebeia),[31] that there is a race of demons who avail themselves of women. Hermes also mentioned this in his *Physika*, and nearly every treatise, both public and esoteric, made mention of this. Thus the ancient and divine scriptures said this, that certain angels lusted after women, and having descended taught them all the works of nature. For this reason they fell into disgrace, he (Hermes?) says, and remained outside heaven, because they taught mankind everything wicked and nothing benefiting the soul. The same scriptures say that from them the giants were born. So theirs is the first teaching concerning these arts [Chemeu]. They called this book *Chēmeu*,[32] whence also the art is called Alchemy (i.e. *chēmeia*)," and so forth.[33]

2. Syriac Zosimus (for a preliminary edition of this text see Appendix)

   Eighth Treatise on the Working of Tin; Letter Ḥēt^h. The Book tells us about tin and Zosimus gives his best greetings to the queen Theosebeia. The holy scriptures say, my lady, that there is a race of demons who has intercourse with the women and has authority over them. Hermes also mentions this story in his *Physika* as well as, so to say, every clear or secret treatise recalls it. In this way,

---

[30] The title *Imouth* is not elsewhere attested in Zosimus's treatises handed down by the Byzantine tradition. Moreover, the reference to the ninth book is not confirmed by the Syriac tradition, where this passage is included in the eighth book by Zosimus. The Arabic *Tome of Images* (*Muṣḥaf aṣ-ṣuwar)*, preserved under the name of the Egyptian alchemist (although its authenticity is questioned; see Hallum, "Tome of Images") preserves a similar account in the sixth book, entitled *Book About the Nature* or *Book of Imouth* (see Abt & Fuad, *Book of Pictures*, 393, 22).

[31] The woman to whom Zosimus usually addresses his treatises; perhaps a pupil of the alchemist: see Hallum, "Theosebeia."

[32] The term *Chēmeu* (*hapax*) is likely to be the title of the book in which the teaching of angels was revealed. The alchemist Olympiodorus (*CAAG* II 80,13) refers to a similar book entitled *βίβλος χημευτική*. Moreover, in the *Corpus alchemicum graecum* several authors mention the alchemist *Chēmēs (Χημής)* or *Chymēs (Χυμής)*, whose name seems to be related to the book *Chēmeu*; see Letrouit, "Alchimistes grecs," 72–4.

[33] Translation by Adler & Tuffin, *George Synkellos*, 18–9 (slightly modified). The Greek text reads: "Ἄξιον δὲ καὶ Ζωσίμου τοῦ Πανοπολίτου φιλοσόφου χρῆσίν τινα παραθέσθαι περὶ αὐτῶν ἐκ τῶν γεγραμμένων αὐτῷ πρὸς Θεοσέβειαν ἐν τῷ ἐνάτῳ τῆς Ἰμοὺθ βίβλῳ, ἔχουσαν ὧδε. 'φάσκουσιν αἱ ἱεραὶ γραφαὶ ἤτοι βίβλοι, ὦ γύναι, ὅτι ἔστι τι δαιμόνων γένος ὃ χρῆται γυναιξίν. ἐμνημόνευσε δὲ καὶ Ἑρμῆς ἐν τοῖς φυσικοῖς, καὶ σχεδὸν ἅπας λόγος φανερὸς καὶ ἀπόκρυφος τοῦτο ἐμνημόνευσε. τοῦτο οὖν ἔφασαν αἱ ἀρχαῖαι καὶ θεῖαι γραφαί, ὅτι ἄγγελοί τινες ἐπεθύμησαν τῶν γυναικῶν καὶ κατελθόντες ἐδίδαξαν αὐτὰς πάντα τὰ τῆς φύσεως ἔργα, ὧν χάριν, φησί, προσκρούσαντες ἔξω τοῦ οὐρανοῦ ἔμειναν, ὅτι πάντα τὰ πονηρὰ καὶ μηδὲν ὠφελοῦντα τὴν ψυχὴν ἐδίδαξαν τοὺς ἀνθρώπους. ἐξ αὐτῶν φάσκουσιν αἱ αὐταὶ γραφαὶ καὶ τοὺς γίγαντας γεγενῆσθαι. ἔστιν οὖν αὐτῶν ἡ πρώτη παράδοσις [Χημεῦ] περὶ τούτων τῶν τεχνῶν. ἐκάλεσαν (ἐκάλεσε in Mosshammer's edition) δὲ ταύτην τὴν βίβλον Χημεῦ, ἔνθεν καὶ ἡ τέχνη χημεία καλεῖται' καὶ τὰ ἑξῆς." I have introduced two changes in Mosshammer's edition. First, I've followed Mertens's suggestion (*Lettre d'Isis*, 67) to expunge the first *Χημεῦ* as an interpolation (the term is not attested in this position by the Syriac tradition). Secondly, in the last line I corrected *ἐκάλεσε* into *ἐκάλεσαν* in accordance with the Syriac translation that ܩܪܘ ܗܘܘ.

in fact, the ancient and holy books tell that some angels fell in love passionately with the women, came down <and> taught them about all the works of nature. For this reason, as the book says, those who acted haughtily remained outside heaven, since they taught mankind all the malicious things which do not benefit the soul at all. The books claim that the works (of nature?) started up from them and theirs is the first exposition concerning these crafts. They called these books *Kwmw*, whence also alchemy (*kumiya*) takes its name. There are 24 treatises in this book. Each of them is given a specific title that is either a letter (of the alphabet) or a word. They are explained by the words of priests. One of these is entitled 'Imus', another 'Imuth', another 'Face'—so it was interpreted (or translated). One of these is entitled 'Key', another 'Seal' or 'Signet', another 'Handbook' (see gr. ἐγχειρίδιον), another 'Position' (of the stars? see gr. ἐποχή). As I said, each one is given a specific title. This book contains the crafts and many thousands of words. Then those who came afterwards, with the intent of doing well, divided the book in many parts; as someone would say: (they did so) in order to compose short versions (of the book) for themselves. And they were not even able to write something useful. For they did not only damage the books of *alchemy*, but also hidden them. The philosopher (i.e. Democritus) claims: 'they hid the writings on the natural substances under the multiplicity of matter.' Perhaps they wanted to exercise our souls. Now, if they exercise the souls, well, philosopher, why to deny it? But you know how to exercise either the body or the soul, and it always leads you to achieve the perfection. In fact a wise saying reads: 'studying is everything.' And also Isidoros says: 'studying increases your work.' I know, this is not beyond your understanding (my lady), but you know it well, since you are one of those who would have liked to hide the art, if it had not been put in writing. For this reason you formed an assembly and administered the oath to each other. But you (my lady) moved away from the various topics (of this book); you presented them in a shorter form and you taught them openly. But you claim that this book cannot be possessed unless in secret. Now, even though secrets are necessary, it is quite fair that anyone has a book of *alchemy*, since it is not kept secret for them. You must know, my lady, he (*i.e.* Democritus?)[34] claimed that those who wrote short versions (of the book) said that just silver can be dyed gold. But the book of *alchemy* they have hidden assured that lead and tin and iron and silver take the color of gold, each metal (takes the color of) the other one, and again the same metals (take the color of) silver, the same metals (the color of) copper, the same metals (the color of) iron. Lead produces tin, copper (produces) iron, silver (produces) gold. In the same way tin

[34] Berthelot & Duval, *Chimie*, 239, considered the philosopher (i.e. ps.-Democritus) as the author of the following quotation. Although the ancient alchemist is cited few lines before (49v18), this identification is not certain.

(produces) the same metals as well. And again from the beginning to the end and from the end to the beginning.

The Syriac text may be divided into two parts. The first section (49^r12–49^v6) matches the Greek passage cited by Syncellus and rephrases the story told by *The Book of Enoch*: in all likelihood Zosimus refers to this book when he mentions the holy scriptures (Sync. 14,3 ἱεραὶ γραφαὶ ἤτοι βίβλοι=SyrZos. 49^r14 ܟܬܒ̈ܐ ܩܘܕ̈ܫܐ). The second section gives a detailed description of the book entitled *Chēmeu* and explains how it was transmitted and somehow corrupted.

In the first part Zosimus attributes the revelation of *chēmeia* both to angels (Sync. 14,6 ἄγγελοί τινες=SyrZos. 49r18 ܥܝܪ̈ܐ) and to demons (Sync. 14,3 δαιμόνων γένος=SyrZos. 49r15 ܓܢܣܐ ܡܕܡ ܕܕܝ̈ܘܐ), so introducing a slight variation into the Enochian myth, where fallen angels usually play the most important role. Although the name of demons is mentioned only at the very beginning of the passage, their presence allows us to read Zosimus's account in the light of the complex (and still unclear in many respects) demonology developed by the author in two other treatises, namely *On the Letter Omega* (*CAAG* II 228–234) and the *First Book of the Final Account* (*CAAG* II 239–245).[35] According to the second work, the two main sources of wealth for Egypt were gold mines and the dyeing techniques, in particular the so-called καιρικαὶ βαφαί (opportune tinctures), that is, the dyeing processes whose success depended on the influence of demons and on the astrological time in which they were performed.[36] This knowledge was considered secret and no ancient authors could reveal it; only Democritus—Zosimus claims—hinted at these tinctures in his explanation of the four τέχναι τιμίαι (valuable arts), which must be identified with the topics covered by the four books, namely the making of gold, of silver, of precious stones, and the purple dyeing of wool.[37] These opportune tinctures were originally called φυσικαὶ βαφαί (natural tinctures) and had been explained by Hermes in his *βίβλος φυσικῶν βαφῶν* (*Book of Natural Dyes*). However, demons became jealous of this knowledge and wanted to make it secret and dependent on their own control (i.e. on the influences of the stars they governed). They started revealing these tinctures, or even a counterfeited form of them (called at some point "unnatural tinctures"), only to their priests in order to be worshipped and receive the appropriate sacrifices.[38]

---

[35] See Fraser, "Zosimos of Panopolis." For a new and reliable edition of the first treatise, see Mertens, *Zosime de Panopolis*, 1–10. The second treatise has been reedited by Festugière, *Révélation d'Hermès*, 363–8 (translation at 275–81).

[36] Festugière, *Révélation d'Hermès*, 363–5 (=*CAAG* II 239–40). See also Mertens, *Zosime de Panopolis*, 62–3 notes 9–10.

[37] Festugière, *Révélation d'Hermès*, 364, ll. 22–4 (= *CAAG* II 242,17), 365, ll. 12–4 (= *CAAG* II 242,8–17). In addition, it is worth mentioning the passage edited in *CAAG* II 242,9–24 (not reedited by Festugière, because 'assez obscure', see 278 note 1), which gives a kind of summary of ps.-Democritus's work by presenting several dyestuffs used by the alchemist as example of καιρικαὶ βαφαί.

[38] Festugière, *Révélation d'Hermès*, 366–7, §§ 6–7 (= *CAAG* II 243–4, §§ 6–7).

Not surprisingly, if we come back to the above-edited passages, Zosimus interpreted Azaēl's revelation in *The Book of Enoch* as referring to the knowledge of tinctures he considered at the basis of any alchemical activity. Τὰ βαφικά are in fact explicitly mentioned by the apocryphal book (see *supra*) along with a specific set of crafts, which is in many respects comparable with the arts explained in ps.-Democritus's four books on dyeing (βαφικαὶ βίβλοι according to Synesius). Moreover, next to *The Book of Enoch* Zosimus mentions also a book of Hermes entitled *Physika* as one of the sources from which he took the account on the angelic/demonic revelation. The identification of this treatise is far from certain, even though the title and the context in which *Physika* is cited remind us of Hermes's *Book of Natural Dyes* quoted in the *First Book of the Final Account*: there Zosimus claimed that Hermes's treatise was addressed to Isidoros,[39] an enigmatic figure that is mentioned also in the Syriac text (50ʳ2).[40]

On the other hand, the Syriac passage emphasises the central role played by books in the revelation and transmission of alchemical knowledge. While *The Book of Enoch* presented Azaēl's revelation in the form of oral teaching, Zosimus underlines the written form in which mankind received this secret knowledge. The arts and all the natural procedures (Sync. 14,7 πάντα τὰ τῆς φύσεως ἔργα=SyrZos. 49ʳ19–20 ܟܠܗܘܢ ܥܒ̈ܕܐ ܕܟܝܢܐ) disclosed by demons had been somehow summarized in the enigmatic books called *Chēmeu*, from which the related term *chēmeia* (χημεῖα or ܟܘܡܝܐ, *kumiya*) derives. However, while the *Final Account* insists on demons' increasing jealousy towards mankind and on their attempt to gain control of the revealed techniques, the Syriac text stresses the point that the original knowledge started deteriorating because of the improper use human beings made of the revealed books. In the same way as the demons (or maybe under their influence) some people tried to hide the books, to summarize them, and disperse their content by focusing only on specific topics.

In this respect, the last part of the Syriac passage is particularly relevant, since Zosimus explains that some alchemists narrowed their inquiry to the methods for gilding silver. Whereas, the art called *chēmeia*, at least in its original form, included a wider set of dyeing techniques that were applied to all kinds of metals in order to dye them different colours. This explanation is consistent with Zosimus's other

[39] Festugière, *Révélation d'Hermès*, 365, ll. 15–20 (= *CAAG* II 242,10–6). The name of Isidoros appears also in the list of alchemists handed down by the *Marcianus gr.* 299 (fol. 7v); see *CAAG* I 110.

[40] We cannot exclude the possibility that other treatises circulating under the name of Hermes referred to or reused the Enochian myth of the fallen angels. In particular, according to a passage of the alchemist Olympiodorus (*CAAG* II 89,9–15), we know that the above-mentioned treatise ascribed to Isis (see *supra*, § 2), or at least some parts of it, was attributed to Hermes. This overlap between Hermes and Isis is not surprising, especially if read in the light of the so-called *Corpus Hermeticum*, which includes several writings where the Egyptian goddess addresses her teaching to Horus (see, for instance, *Stobei fragmenta*, xxiii–xxvii). Therefore, Zosimus could have had in mind Isis's book, when he mentioned Hermes's *Physika*: see Scott, *Hermetica*, 151.

alchemical books that have been preserved by the Syriac manuscript Mm. 6.29. In fact, among the twelve treatises handed down by the codex under the name of the Egyptian alchemist, we find many writings devoted to the dyeing of various metals, such as silver (e.g. Book 2; Berthelot-Duval, *La chimie*, 217–21), tin (Book 8; *idem*, 238–42), lead (Book 10; *idem*, 253–57), and iron (Book 11; *idem*, 257–60).[41] In particular Book 6 (Mm. 6.29, fol. 32x^{v}17–45^{r}8), entitled *Beginning of the Treatise on the Working of Copper: Letter Waw*, includes several recipes explaining how to process copper and dye it black, purple, coral red, white, and yellow (Berthelot-Duval, *La chimie*, 222–32). Scholars have so far focused their attention especially on the recipes dealing with the production of black copper alloys. Three recipes have been recently published by Erika Hunter and fully commented by Alessandra Giumlia-Mair, who recognized in them the description of different methods for producing a black and shiny patina on copper-alloys.[42] This black copper was already known in Ancient Egypt, as we can infer, for instance, from a beautiful black image of the pharaoh Amenemhat III (1842–1794 BC) today displayed at the Ortiz Collection in Geneva.[43] It represents one of the most ancient examples of artificially black patinated statues found in Egypt (Fig. 1).

Conclusion

If we go back to the above-quoted Syriac passage, in the very last part Zosimus criticizes the attitude of those alchemists who were only interested in the making of precious metals by quoting a sentence presumably taken from an ancient author (perhaps ps.-Democritus) who claimed: "those who wrote short versions [of the books of *chēmeia*] said that just silver can be dyed gold" (50^{r}12–3). In this way the debate between people supporting a wider idea of alchemy, which included a broader set of dyeing techniques, and people accepting a narrower idea just focused on the making of gold seems to be traced back to the earliest phases of alchemy in Egypt. A possible target of such a criticism might be recognized in Isis's treatise, a work that mentioned only the making of gold and silver among the secrets revealed by the angels. On the contrary, a different position was endorsed by ps.-Democritus with the fourfold division of his books. Significantly, as we have already noted, the

[41] The Syriac text of these twelve treatises is still unedited (an edition with English translation is scheduled to be published within the new series 'Sources of alchemy and chemistry' distributed as a supplement of *Ambix*); a partial French translation is available in Berthelot & Duval, *Chimie*, 210–66.

[42] Hunter, "Beautiful Black Bronzes," 656–7. The three recipes that have been edited correspond with the texts translated or summarized by Berthelot & Duval, *Chimie*, 223 (rec. 2), 224–5 (rec. 8–9), 225 (rec. 12); and Giumlia-Mair, "Zosimos the Alchemist," 319–21.

[43] See, for example, Giumlia-Mair, "Krokodil."

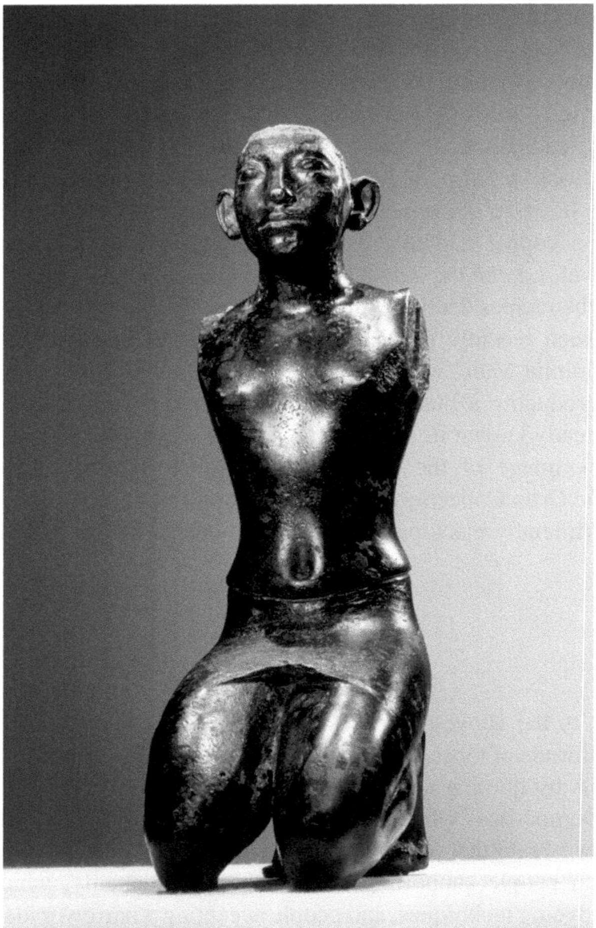

Fig. 1 Black copper statue of King Amenemhat III in kneeling posture (1843–1798 BC) (Courtesy of Werner Forman Archive, The Bridgeman Art Library)

chemical Papyri seem to preserve a similar division, which should lead us to speculate on how widespread or how standard such a division might actually have been in the first centuries AD.

Such a background was well known to Zosimus, who took for granted that *chēmeia* was related to a plurality of dyeing techniques. The proper alchemist, Zosimus claims, was expected to put into practice any possible transformation of colour without focusing only on gold. Nevertheless, the Egyptian alchemist, though very familiar with ps.-Democritus's books, never mentions stones working and purple dyeing in the passage. These kinds of crafts seem to have been somehow left apart within the debate on the contents of the original books of alchemy. Although Zosimus stresses the central role played by dyeing techniques, the polarity between a wider idea of alchemy and the bare making of precious metals seems to have been reduced to a simpler discussion on metalworking: Zosimus underscores all the possible chromatic transformations of metals, which seem to become the most relevant topic of any alchemical inquiry.

Such a variety of positions already detectable during the earliest phases of the western alchemical tradition makes clear how fluid were the boundaries of an art primarily focusing on the chromatic transformations of the treated substances. If the Byzantine sources analysed at the beginning of the chapter emphasizes the identification of *chēmeia* with the making of gold and silver, such a position—although dominant to some extent—never interrupted the transmission of technical knowledge dealing with a wider set of dyeing techniques. A similar kind of expertise, for instance, is well attested by the Latin recipe books of the Middle Age, which covered a variety of topics more similar to ps.-Democritus's four books and to the Leiden and Stockholm papyri than to the simple selection of recipes made by Psellos.[44] A well-known example is the so-called *Mappae clavicula*, a large collection of recipes probably assembled between the ninth and the twelfth century. Recent studies have shown that this work in all likelihood derives from a lost Greek source that could be perhaps identified, according to Halleux-Meyvaert's investigation, with a lost part of Zosimus's treatises.[45] Although the extant sources do not allow us to confirm such a hypothesis, the Greek origin of the *Mappae clavicula* confirms the transmission of a set of practices originally covered by those texts considered at the basis of the Byzantine alchemical tradition. To sum up, the mere practice of *chrysopoeia* does not seem to be sufficient for getting a full understanding of the different historical evolutions of a complex art, whose exact definition was debated even among the authors usually considered as the founders of this discipline. On the contrary, a deeper investigation of their treatises—many of them still waiting for editions and translations—provide us with new and fresh material for building a more complete picture of the evolution and transformation of the different technical aspects somehow encapsulated in the enigmatic word *chēmeia*.

Appendix

Zosimus's passage on the relevation of alchemy as preserved by the Syriac manuscript Mm. 6.29, fols. 49^{r}–50^{r}:[46]

[44] See, for instance, Halleux, *Papyrus de Leyde*, 53–62.

[45] Halleux & Meyvaert, "Mappae clavicula," 12–3.

[46] A French translation of the passage is available in Berthelot & Duval, *Chimie*, 238–9.

[49ʳ12-20]

ܠܘܬ ܡܠܐܟܐ ܕܐܡܝܢܐ ܕܒܠܐ ܥܒܕܐ ܕܪܘܫ: ܚܕ ܐܬܘܬܐ:
ܟܠܗܐ ܕܒܠܐ ܙܘܓ ܡܫܬܠܛܐ ܠܗ ܘܐܘܣܝܐܣ: ܠܗܘܬ
ܠܐܘܣܝܐ ܡܬܒܥܝܐ ܦܣܝܣ ܦܟܝܢ. ܟܠܗ̈ܝܢ ܒܢܝ̈ܐ ܐܘ̈ ܐܢܫ̈ܐ ܐܡܪܝܢ
ܕܐܝܬܘ̈ ܒܢܫܐ ܡܕܡ ܕܙܒܢܐ. ܗܘ̈ ܕܡܬܐܡܢ ܒܢ̈ܫܐ ܘܡܕܒܪ
ܠܗܘܢ. ܚܕ ܕܝܢ ܐܦ ܐܪܣܛܘ ܒܒܢܝܢܫܐ. ܘܐܝܟ ܕܝܢ
ܕܒܡܐܡܪ ܒܟܠܗ ܡܠܐܡܪܐ ܒܟܠܐ ܘܡܬܩܢܐ. ܒܟܬܒܐ ܡܟܝܠ.
ܗܟܝܠ ܒܡܢ ܐܡܪܝܢ ܟܬܒ̈ܐ ܦܪ̈ܝܫܐ ܘܐܠܗ̈ܐ. ܘܡܢ ܩ̈ܕܡܝܐ
ܠܢ ܐܬܝܗܒܘ ܦܠܛܐ ܕܒܠܐ ܢܦ̈ܫܐ ܘܫܠܘ <ܘ>ܐܠܗܘ ܐܢܫ ܟܠܗܘ<ܢ>
ܥ̈ܒܕܐ ܕܟܝܢܐ. ܗܟܝܠ ܕܡܬܦܠܓܝܢ ܐܝܟ ܕܐܡܪ ܟܠܗܐ ܕܐܝܬܝܢ

[49ᵛ1-21]

ܕܐܬܟܬܒܘ ܡܢ ܟܬܪ ܡܢ ܩܡܝܐ ܦܬܘ. ܡܛܠ ܕܟܠܗܘܢ
ܫܢܝ̈ܐ ܕܠܐ ܡܬܘܪ̈ܩܢ ܡܕܡ ܟܢܘܫܝܐ ܐܠܗܘ ܒܒܢܝ̈ܢܫܐ.
ܡܢܗܘܢ ܗܟܝܠ ܐܡܪܝܢ ܩܠܗܐ. ܗܟܝܢ ܕܐܬܥܒܕ ܥ̈ܒܕܐ.
ܘܐܝܬܘ ܡܫܬܡܫܢܘܬܐ ܦܪܫܐ ܕܡܢܗܘܢ. ܟܠ ܗܟܝܢ
ܐܘܡܢ̈ܘܬܐ. ܩܪܝܢ ܗܘܘ ܟܬ̈ܒܐ[47] ܕܟܘ̈ܡܘ. ܘܡܢ ܗܪܟܐ ܘܐܦ
ܟܘܡܐ ܡܬܐܡܪܐ. ܐܝܬ ܒܗ ܕܝܢ ܒܟܬܒܐ ܒܫܢ̈ܝܢ ܘܐܪ̈ܒܥܐ
ܛܘ̈ܦܣܐ. ܘܟܬܒܝܢ ܕܝܢ ܡܢܗܘܢ ܫܡܐ ܕܢܦܫܗ ܐܝܬ ܒܗ.
ܐܘ ܐܬܘܬܐ ܐܘ ܡܠܐܡܪܐ. ܘܡܬܦܫܩܝܢ ܒܟܬܒܐ ܕܒܘܡܪ̈ܐ.
ܚܕ ܡܢܗܘܢ ܕܝܢ ܡܬܐܡܪ ܐܡܘܣ. ܐܚܪܢܐ ܐܡܘܢܐ. ܐܚܪܢܐ
ܐܦ̈ܐ. ܡܬܦܫܦ ܕܝܢ ܗܟܢܐ. ܚܕ ܡܢܗܘܢ ܡܬܐܡܪ ܡܟܝܪܐ.
ܘܐܚܪܢܐ ܕܝܢ ܛܒܥܐ ܐܘ ܣܠܡܝܐ. ܐܚܪܢܐ ܐܝܟܢܘܡܘܢ. ܐܚܪܢܐ ܐܦܘܟܐ.
ܘܐܝܟ ܕܐܦܪܐ ܟܟܝܢ ܡܢܗܘܢ ܫܡܐ ܕܟܠܗ ܐܝܬ ܒܗ. ܐܝܬ ܒܗ
ܕܝܢ ܒܟܬܒܐ ܗܢܐ. ܐܘܩܝܢܘܬܐ ܘܐܠ̈ܦܐ ܫܪ̈ܒܐ ܕܩܠܐ. ܗܟܝܢ
ܗܟܢܐ ܕܡܢ ܟܠܪܘܗܝ ܗܘܝ: ܐܝܟ ܗܘ ܕܛܥܐ ܥܒܕܝܢ ܗܘܘ [?]

47 The sense of the sentence, in which ܟܬ̈ܒܐ ܕܟܘ̈ܡܘ (the books of *kwmw*) seems to be the object of ܩܪܝܢ ܗܘܘ (they called), is difficult; perhaps ܟܬܒܐ must be supplied before ܕܟܘ̈ܡܘ. In addition the plural form 'the books' does not correspond with the Greek version that has a singular. The plural might be justified in the light of the following part, according to which the book revealed by demons was divided into several tomes.

ܦܠܓܘܗܝ. ܘܐܝܟ ܐܝܬ ܕܐܡܪ. ܐܝܟ ܕܒܢܦܫܗܘܢ ܦܫܝܛܬܐ
ܒܟܝܢ (?)[48] ܗܘܘ. ܐܦܠܐ ܕܝܢ ܗܢܘܢ ܗܠܝܢ ܡܕܡ ܕܨܒܥ ܐܠܓܒܗ.
ܐܠܐ ܠܐ ܗܘܝܢ (?) ܒܟܝܢܗܘܢ ܗܢܘܢ ܒܟܠܗܝܢ ܕܓܘܫܡܐ. ܐܠܐ ܐܦ ܓܢܣܘ.
ܐܡܪ ܕܝܢ ܦܝܠܘܣܘܦܐ. ܕܒܫܘܓܐ ܘܡܟܘܬܐ ܓܒܪܘ ܐܢܫ
ܒܚܬܝܬܐ ܡܢܝܬܐ. ܒܬܪ ܕܝܢ ܕܒܢܦܫܬܗܘܢ ܙܒܢ ܕܢܘܪܗܘܢ. ܐܝ
ܗܟܢܐ ܒܢܦܫܬܐ ܡܕܘܪܝܢ: ܐܝܟ ܦܝܠܘܣܘܦܐ ܒܟܡܐ ܐܝܬ
ܡܕܒܚܐ. ܐܠܐ ܢܕܥ ܕܕܘܪܐ ܗܘ. ܐܘ ܕܒܘܡܝܐ ܗܘ. ܐܘ ܕܢܦܫܐ

[50r1-18]

ܗܘ. ܘܒܒܠܒܝܢ ܠܥܘܡܩܐ ܡܢܝܬܐ. ܐܡܪܐ ܓܝܪ ܗܘ ܡܠܟܬܐ
ܣܝܡܬܐ. ܘܒܠܡܕܝܢ ܗܪܓܐ ܗܘ. ܘܐܦ ܐܣܝܕܘܣ (sic) ܐܡܪ.
ܕܗܪܓܐ ܒܚܕܐ ܡܫܓܐ. ܢܕܥ ܐܢܐ ܕܐܦܠܐ ܚܒܪܐ ܒܟܝ. ܐܠܐ ܐܦ
ܐܢܬܝ ܢܕܥܐ ܐܢܬܝ. ܕܚܕ ܡܢ ܗܠܝܢ ܐܢܬܝܢ ܕܙܒܢ ܕܢܦܫܘܢ
ܐܘܡܢܘܬܐ. ܠܗ ܠܐܒܘܗ. ܡܬܚܫܒܐ ܐܦ ܒܢܘܫܐ ܒܓܒܪܘܢ.
ܘܡܩܘܡܬܐ ܕܒܗܝ ܚܕܪܐ ܫܡܫܘܢ. ܐܠܐ ܐܢܬܝ ܦܪܫܬܝ ܡܢ
ܩܠܘܬܐ ܫܟܝܚܐ. ܘܐܢܬܝ ܐܝܟ ܠܡܠܘܬܐ ܕܒܘܪܐ. ܘܠܘܩܒܠܝ
ܐܝܟ ܓܕܫܐܝܬ. ܐܠܐ ܐܡܪܐ ܐܢܬܝ ܕܠܐ ܡܬܩܢܝܐ ܟܠܟܠ ܗܝ ܐܠܐ ܒܪܐܙܐ. ܐܝ
ܗܘ ܗܟܢܐ ܕܙܐܙܐ ܡܬܟܒܝܢ ܗܘܘ: ܠܬܪܐܝܬ ܕܝܢ ܗܘܐ ܕܒܟܒܝܢ
ܗܘܐ ܟܠܟܠ ܕܒܘܡܝ:[49] ܟܕ ܠܐ ܡܬܓܫܡܐ ܗܘܐ ܡܢܗܘܢ. ܕܟܝ

[48] The word is indecipherable; see *infra*, fol. 50r11 ܕܒܫܡܝܢܬܐ ܡܬܓܒܠܘ.
[49] The term ܒܘܡܝ (see also 50r12) is likely to be a slightly different spelling of ܒܘܡܝܐ.

ܘܟܠ ܐܘ ܐܢܬܬܐ ܕܐܡܪ ܕܗܟܢ ܕܝܢ ܕܒܦܢܩܝܬܐ ܟܬܒܘ: ܕܡܢ
ܗܘܪܐ ܐܡܪ ܒܟܬܒܗ ܕܡܬܩܪܐ ܫܡܫܐ. ܒܐܬܐ ܕܝܢ ܕܒܗܘܢ
ܗܘ ܕܫܡܫ ܡܫܡܠܝܐ: ܕܐܝܟ ܙܪܥ ܘܐܪܥܐ ܘܫܡܫܐ ܙܒܢܝܢ
ܫܡܫܐ. ܒܟܝܢ ܡܢܗ ܘܒܗ. ܠܘܬ ܕܝܢ ܗܘܝ ܒ ܗܘܝ
ܗܘܪܐ. ܘܠܘܬ ܗܘܝ ܒ ܗܘܝ ܒܟܠܐ. ܘܗܘܝ ܒ ܗܘܝ ܐܪܥܐ.
ܒܐܝܟ ܕܝܢ ܟܒ ܙܘܓܐ. ܘܒܟܠܐ ܐܪܥܐ. ܘܫܡܫܐ ܫܡܫܐ. ܗܟܢܐ
ܐܦ ܙܘܓܐ ܗܘܝ ܒ ܗܘܝ. ܐܦ ܗܟܢܐ ܡܢ ܟܠܗ ܚܠܝܬܐ. ܘܡܢ
ܚܠܝܬܐ ܟܠܗ.

Bibliography

Abt, Theodor, and Salwa Fuad. 2011. *The Book of Pictures by Zosimos of Panopolis*. Zurich: Living Human Heritage Publication.

Adler, William, and Paul Tuffin. 2002. *The Chronography of George Synkellos. A Byzantine Chronicle of Universal History from the Creation*. Oxford: Oxford University Press.

Albini, Francesca. 1988. *Michele Psello. La Crisopoea, ovvero come fabbricare l'oro*. Genova: ECIG.

Baar, James. 1979. Aramaic-Greek Notes on the Book of Enoch (ii). *Journal of Semitic Studies* 24: 179–192.

Berthelot, Marcelin, and Charles-Émile Ruelle. 1887–1888. *La Collection des anciens alchimistes grecs*, 3 vols. Paris: G. Steinheil.

Berthelot, Marcelin, and Rubens Duval. 1893. *Histoire des Sciences: La Chimie au Moyen-Âge*, vol. 2: *L'alchimie syriaque*. Paris: Imprimerie nationale.

Bidez, Joseph. 1928. *Michel Psellus. Épitre sur la Chrysopée, opuscules et extraits sur l'alchimie, la météorologie et la démonologie* (*Catalogue des manuscrits alchimiques grecs*, vol. 6). Brussels: M. Lamertin.

Bidez, Joseph, and Franz Cumont. 1938. *Les Mages hellénisés. Zoroastre, Ostanès et Hystaspe d'après la tradition grecque*, 2 vols. Paris: Les Belles Lettres.

Black, Mathew. 1985. *The Book of Enoch or 1 Enoch: A New English Edition with Commentary and Textual Notes*. Leiden: Brill.

Brooks, Ernest W. 1900. A Syriac Fragment. *Zeitschrift der Deutschen Morgenländischen Gesellschaft* 3: 195–230.

Caley, Earle R. 1926. The Leyden Papyrus X: An English Translation with Brief Notes. *Journal of Chemical Education* 3: 1149–1166.

Caley, Earle R. 1927. The Stockholm Papyrus: An English Translation with Brief Notes. *Journal of Chemical Education* 4: 979–1002.

Chabot, Jean-Baptiste. 1910. *Chronique de Michel le Syrien*, vol. 4: *Texte syriaque*. Paris: Pierre Leroux.

Chabot, Jean-Baptiste. 1933. *Incerti auctoris Chronicum Pseudo-Dionysianum vulgo dictum*, vol. 2 (*CSCO* 104, syr. 53). Louvain: Peeters.

Dodge, Bayard. 1970. *Abū 'l-Faraj Muḥammad ibn Isḥāq al-Nadīm, The Fihrist, a 10th Century AD Survey of Islamic Culture*. New York: Columbia University Press.

Duffy, John M. 1992. *Michaelis Pselli philosophica minora*. Leipzig: Teubner.

Duval, Rubens. 1888–1901. *Lexicon Syriacum auctore Hassano bar Bahlule*, 3 vols. Paris: E. Reipublicae typographaeo.

Festugière, André-Jean. 1950. *La Révélation d'Hermès Trismégiste*, vol. 1: *L'Astrologie et les sciences occultes*. Paris: J. Gabalda et Cie.

Flügel, Gustav. 1872. *Kitāb al-Fihrist*, 2 vols. Leipzig: F.C.W. Vogel.

Fraser, Kyle A. 2004. Zosimos of Panopolis and the Book of Enoch: Alchemy as Forbidden Knowledge. *Aries* 4: 125–147.

Fück, Johann W. 1951. The Arabic Literature of Alchemy According to al-Nadīm (A.D. 987). A Translation of the Tenth Discourse of the Book of the Catalogue (*Al-Fihrist*) with Introduction and Commentary. *Ambix* 4: 81–144.

Gildemeister, Johannes. 1876. Alchymie. *ZDMG* 30: 534–538.

Giumlia-Mair, Alessandra. 1996. Das Krokodil und Amenemhat III. aus el-Faiyum. *Antike Welt* 27: 313–321.

Giumlia-Mair, Alessandra. 2002. Zosimos the Alchemist – Manuscript 6.29, Cambridge, Metallurgical Interpretation. In *I bronzi antichi: produzione e tecnologia* (Atti del XV Congresso Internazionale sui Bronzi Antichi, organizzato dall'Università di Udine, sede di Gorizia, Grado-Aquileia, 22–26 maggio 2001), ed. Alessandra Giumlia-Mair, 317–323. Montagnac: Éditions Monique Mergoil.

Goeje, Michael Jean de. 1885. *Ibn al-Faqīn al-Hamaḏānī, Mukhtaṣar kitāb al-buldān*. Leiden: Bibliotheca Geographorum Arabicorum.

Halleux, Robert. 1981. *Alchimistes grecs*, vol. 1: *Papyrus de Leyde. Papyrus de Stockholm. Fragments de recettes*. Paris: Les Belles Lettres.

Halleux, Robert, and Paul Meyvaert. 1987. Les Origines de la Mappae clavicula. *Archives d'Histoire Doctrinale et Littéraire du Moyen-Age* 54: 7–58.

Hallum, Bink. 2008. Theosebeia. In *Encyclopedia of Ancient Natural Scientists*, eds. Paul Keyser and Georgia Irby-Massie, 302–303. London/New York: Routledge.

Hallum, Bink. 2009. The Tome of Images: An Arabic Compilation of Texts by Zosimus of Panopolis and a Source of the *Turba Philosophorum*. *Ambix* 56: 76–88.

Hunter, Erika C.D. 2002. Beautiful Black Bronzes. Zosimos' Treatises in Cam. Mm. 6.29. In *I bronzi antichi: produzione e tecnologia* (Atti del XV Congresso Internazionale sui Bronzi Antichi, organizzato dall'Università di Udine, sede di Gorizia, Grado-Aquileia, 22–26 maggio 2001), ed. Alessandra Giumlia-Mair, 655–660. Montagnac: éditions Monique Mergoil.

Katsiampoura, Gianna. 2008. Transmutation of Matter in Byzantium: The Case of Michael Psellos, the Alchemist. *Science & Education* 17: 663–668.

Knibb, Michael A. 2009. *Essays on the Book of Enoch and Other Early Jewish Texts and Traditions*. Leiden: Brill.

Letrouit, Jean. 1995. Chronologie des alchimistes grecs. In *Alchimie: art, histoire et mythe* (Actes du 1er Colloque International de la Société d'Étude de l'Histoire de l'Alchimie), eds. Didier Kahn and Sylvain Matton, 11–93. Paris/Milan: S.É.H.A.- Arché.

Martelli, Matteo. 2011. *Pseudo-Democrito. Scritti alchemici con il commentario di Sinesio*. Paris/ Milan: S.É.H.A.-Archè.

Massé, Henry. 1973. *Ibn al-Faqīn al-Hamaḏānī. Abrégé du livre de Pays*. Damascus.

Mertens, Michèle. 1983–1984. *Un traité gréco-égyptien d'alchimie: la lettre d'Isis à Horus. Texte établi et traduit avec introduction et notes*. Université de Liège: Mémoire de licence.

Mertens, Michèle. 1995. *Zosime de Panopolis. Mémoires authentiques*. Paris: Les Belles Lettres.

Mertens, Michèle. 2006. Greco-Egyptian Alchemy in Byzantium. In *The Occult Sciences in Byzantium*, eds. Paul Magdalino and Maria Mawroudi, 205–229. Geneva: La Pomme d'Or.

Milik, Jazef T. 1976. *The Books of Enoch, Aramaic Fragments of Qumrân Cave 4*. Oxford: Clarendon.

Reitzenstein, Richard. 1904. *Poimandres. Studien zur griechsch-ägyptischen und frühchristlichen Literatur*. Leipzig: B.G. Teubner.

Roberto, Umberto. 2005. *Ioannis Antiocheni Fragmenta ex Historia Chronica*. Berlin: De Gruyter.

Scott, Walter. 1985. *Hermetica*, vol. 4: *Testimonia with Introduction, Addenda and Indices by. A.S. Ferguson*. Boston: Shambhala.
Strohmaier, Gotthard. 1991. Umāra ibn Ḥamza, Constantine V and the Invention of the Elixir. *Graeco-Arabica* 4: 21–24.
Westerink, Leendert G. 1948. *Michael Psellus, De Omnifaria Doctrina*. Nijmegen: Centrale Drukkerij N.V.

Transmission of Alchemical and Artistic Knowledge in German Mediaeval and Premodern Recipe Books

Sylvie Neven

Abstract In the Middle Ages and premodern period, artisanal knowledge was transmitted via collections of recipes often grouped concomitantly with alchemical texts and instructions. Except for some very well-known artistic treatises, e.g. works by Eraclius or the *Schedula diversarum artium*, attributed to Theophilus, detection and delimitation of alchemical content within recipe books has been rare and fraught with difficulty. Alchemy can be defined as the 'art of transmutation', referring to the perfection of base or impure matter (often metal or stone) into perfect substances. Alchemical procedures thus rely on artisanal/craft practices. Any overlap between alchemy and art-technological procedures can be explained by the use of identical materials and substances. Both are concerned with the description of colours—especially in processes of change, the making of pigments, the production of artificial gemstones, the imitation of gold and silver and the transmutation of materials. Both require procedures involving precise and specifically defined actions, prescriptions and ingredients. So both ultimately use identical rhetorical formulations that reflect a 'step by step' procedure. Assuming that alchemical and artistic texts have the same textual format, raises the question: did they also have the same types of production and dissemination? Using a corpus of about 40 manuscripts produced in Northern Europe between the fourteenth and the sixteenth centuries, this paper investigates the context behind these writings, and the various ways alchemical and artisanal recipes were embedded within recipe books. It also proposes some clues to assist in locating, identifying and demarcating alchemical writings within the literature of recipes.

In the Middle Ages and premodern period, alchemical knowledge and practice was frequently transmitted via collections of recipes grouped concomitantly with artistic instructions. Presented in the form of a succession of more or less short notes, these

S. Neven (✉)
FRS-F.N.R.S, University of Liège (ULg), 'Transitions' Liège, Belgium
e-mail: Sylvie.Neven@ulg.ac.be

S. Dupré (ed.), *Laboratories of Art*, Archimedes 37,
DOI 10.1007/978-3-319-05065-2_2,

writings describe processes for the manufacture, preparation and application of various types of materials and substances. The majority are anonymous compilations of texts, which may originate from older or undetermined authorities. Hundreds of such collections of recipes dealing both with alchemical and art-technological procedures were produced and disseminated in Northern Europe from the fourteenth century on, especially in German-speaking countries.

Drawing on a delimited corpus of about 40 representative German manuscripts dated from the fourteenth to the sixteenth century, this paper investigates the connections and similarities between these two fields and examines the various ways alchemical and artistic instructions were embedded within recipe books.[1] It argues that textual form and lexical proximities within recipes from these different disciplines may lead to association, contamination and confusion within this textual genre. It finally suggests some clues to help locate, distinguish and demarcate alchemical content within the literature of recipes.

Art and Alchemy Within Recipe Books

At first sight, any overlap between alchemy and art-technology within recipe books can be broadly explained by the mutual use of various materials and substances such as "common stones, gems, and types of marble, gold and other metals, sulfurs, salts, and inks, azures, minium, and other colors, oils and burning pitches, and countless other things."[2] More precisely, the field of art-technology encompasses a large range of craft practices involved in the production of pieces of art (including those which incorporate such substances). This 'hand' knowledge, is related to the mechanical arts and is divorced from the philosophical or speculative dimension. Yet, alchemy could be described as the practical, philosophical and medical search for the perfecting of base material substances and also for the extending of life.[3] The theoretical and practical aspects of alchemy involve both the study of all inanimate or animate things made from the elements and the observation and imitation of natural processes within the laboratory.[4] In this context, alchemy could be seen as a mechanical art, in the sense that it works on matter but is also a liberal art, as it attempts to explain matter in its composition and its transformation.[5] In the practical sense, one of the main goals of alchemy is the transmutation

---

[1] The main data and characteristics of these manuscripts are given in Appendix.

[2] Brewer, *Fr. Rogeri Bacon*, 39–40.

[3] Perception and definition of alchemy is not chronologically constant and has been the subject of several (re)interpretations since the eighteenth century, see Principe & Newman, "Historiography of Alchemy."

[4] Halleux, "Alchimie," 336–7; Newman, "Technology and Alchemical Debate," 432–3; Pereira, "Use of Vernacular Languages," 336; and Kahn, *Alchimie et paracelsisme*, 7–8.

[5] Halleux, *Savoir de la main*, 134.

of base or impure matter (often metal or stone) into a noble or perfect one.[6] To do so, alchemists used to perform chemical processes and manipulations which resembled those practiced by contemporary artists and artisans.

Thus, in both fields particular importance is placed on craft practices. Both alchemical and artistic recipe books describe various processes for purifying and transforming materials, either for improving their properties or in order to use them for specific purposes. In this context, from a technological point of view, the term 'transmutation' could refer to the colouring of glass, the melting and tinting of metals, the dyeing of stones or gems, or the manufacture of synthetic pigments. A huge number of recipes are dedicated to procedures for obtaining gold or silver or gold-silver like substances from base metals (copper, tin, lead, iron, mercury).

The tradition of recipe books has roots deep in Antiquity. Treatises dealing both with art-technological and alchemical procedures notably survived in the Alexandrian Greek papyri preserved in Leyden and Stockholm.[7] These date from the third century but were probably based on previous texts.[8] These papyri contain information on the imitation of precious materials such as gold, gems and Tyrian purple. They also have recipes dedicated to the art of dyeing, to chrysography, and to the making of artificial precious stones.

An echo of these recipes can be found in the Codex Lucensis 490 (Lucca, Biblioteca Capitolare Feliniàna), also known as the *Compositiones ad tingenda musiva* or *Compositiones variae*. This manuscript, dated from the end of the eighth or the very beginning of the ninth century, is based on previous Greek sources compiled around the beginning of the seventh century.[9] This recipe book deals with various artistic techniques, notably the dyeing of skins, the manufacture of pigments, colorants, varnishes and glues, chrysography and the gilding of metals.

The *Lucca Manuscript* shares content with the *Mappae clavicula*, compiled around 800. Parts of this text have far earlier origins and present parallels not only with the Leyden and Stockholm papyri but also with Syriac translation of Zosimus and with ps.–Democritus's writings.[10] The nucleus of this text was probably a Greek alchemical treatise written and translated between the fourth and the fifth centuries, with additions in the eleventh and twelfth centuries.[11] The two main manuscripts are the Corning, Corning Glass Museum, Ms. 5 and Sélestat, Bibliothèque Humaniste, Ms. 17, but this tradition was also (partly) disseminated through dozens of manuscripts. The *Mappae clavicula* contains about 300 recipes

[6] Singer, *Catalogue of Latin*, 38; Principe & DeWitt, *Transmutations*, 2–3; and Principe, *Secrets of Alchemy*, 13.

[7] On Leyden, see Rijksmuseum van Oudheden, Papyrus P. LEID.X.; On Uppsala, Victoria Museum, P. HOLM.

[8] Halleux, *Papyrus de Leyde*.

[9] Hedfors, *Compositiones ad tingenda musiva*; and Johnson, *Compositiones variae*.

[10] Berthelot & Ruelle, *Anciens alchimistes grecs*. See also Martelli, *Pseudo–Democrito*.

[11] Halleux & Meyvaert, "Mappae clavicula"; and Berthelot & Duval, *Chimie au Moyen-Âge*, vol. I.

and descriptions of miscellaneous chemical operations, including instructions for the manufacture of dyes and pigments, for the gilding and painting on glass, as well as, among others, metalwork, chrysography, distilling alcohol, making candy, and creating military devices.[12]

Mediaeval and Premodern Recipe Books

In mediaeval and premodern times, artistic and alchemical procedures were often described within compilations of texts that may concurrently address various fields such as medicine, cooking, botany or pharmacology. They also include magical recipes, dietetical instructions or advice on home-economics. All these various disciplines are embedded within the genre of the *Fachliteratur*.[13] This kind of literature regroups all texts of a utilitarian and informative nature whose content does not principally concern aesthetic or religious issues, or matters relating to emotional purpose.[14] A great number of these writings share the same format and are quite similar in terms of their external and internal characteristics. Within these compilations, the recipe frequently appears as the "shortest element in which the text could ultimately be divided."[15] This observation, although initially relating to the field of alchemy, can also be applied to recipe books in general during the Middle Ages and the premodern period. Robert Halleux underlined the similarity in format between the mediaeval treatises of alchemy and the so-called technical recipe books. He states that, whatever subject the recipe books are dedicated to, they all present a similar structure, from the earliest Mesopotamian examples to the pharmacopoeia texts of the sixteenth century. We could refine this definition by adding that the recipe is the smallest 'independent' element into which these texts could be divided. In fact, a recipe could be seen as an independent text in itself and could thus be dissociated from its original recipe book and be introduced into the pages of another manuscript. For this reason, it may be argued that the recipe, as a type text, could be considered as a structural unit common to several disciplines embedded within the manuscripts belonging to the *Fachliteratur* and serve to define a genre in itself. As Bruno Laurioux noted "[the recipe] gives the tone and standardizes, by its repetitive structure, the corpus of this literary genre."[16]

---

[12] Smith & Hawthorne, "Mappae Clavicula."

[13] The *Fachliteratur* has been the subject of various studies. Concerning the German production, see notably Eis, *Mittelalterliche Fachliteratur*, and more recently Haage & Wegner, *Fachliteratur der Artes*.

[14] Jansen–Sieben, *Repertorium*, XII.

[15] Halleux, *Textes alchimiques*, 74.

[16] "C'est elle qui donne le ton et uniformise, par sa structure répétitive, l'ensemble de ce genre littéraire." (Laurioux, *Livres de cuisine médiévaux*, 13).

Craft practices, alchemical treatises and artists' recipe books thus share parts of the same specific syntax, the frequent use of the imperative form but also some particular verbs (such as 'grind', 'mix', 'purify') and vocabulary. For example, the first alchemists used the word *tinctura* to refer to the tinting or the dyeing of metals, stones or clothes.[17] These methods notably explained how to dye metals yellow or white—so (apparently) how to transmute them into gold or silver. They also described various ways to counterfeit precious stones.[18] In this context, the term *tinctura* does not relate to the artisanal practice of dyeing, but instead describes the procedure for executing the transition from one colour to another, through the steps of the alchemical process. Another example is provided by the terms 'mercury' and 'sulfur'. According to the context, these may alternatively designate the common substances used for making vermilion or the two principles of which all metals were thought to be composed in different proportions.[19]

Thus, both artistic practices and alchemy required procedures involving precise and specifically defined actions, prescriptions and ingredients. So both used an identical rhetorical recipe formulation that reflects a 'step by step' procedure.

Assuming that alchemical and artistic texts have the same format and were assembled within the same sort of compilation raises the question: were they produced, diffused and read by the same people? Previous research has demonstrated that investigating questions related to the authorship and the context of production behind these texts, as well as their compilation and dissemination, elucidates information about the former nature and the previous and current function of these writings.[20] Answering these questions would: first, help to better estimate the relevance of these books when using them as a historical source for reconstructing part of mediaeval and premodern alchemical and artistic knowledge. And second, examining the various connections and similarities between these two fields, as described within recipe books, would serve to (re)situate them in their historical and cultural contexts.

The Sources and the Context of Production

First of all, the wide diversity of subjects and fields embedded within the corpus begs the question: were they written by several authors? *A priori*, palaeographical examination tends to confirm this: as with a large number of recipe books produced during mediaeval and premodern times, the manuscripts examined were written by several

[17] Principe, *Secrets of Alchemy*, 17; and Clarke, *Art of All Colours*, 37.

[18] See in this volume Matteo Martelli.

[19] Bucklow, "Paradigms and Pigment Recipes"; and Principe, *Secrets of Alchemy*, 35–6.

[20] Neven, *Recettes artistiques*.

hands, and these hands are predominately anonymous. These works thus appear to be the result of collaboration, or at least intervention, by several distinct persons. However, each person's contribution cannot necessarily be allocated according to the different subjects in the book. The same hand might be responsible for both a medical treatise and a collection of alchemical or art technological recipes.

The manuscripts under consideration were, in fact, the result of copying and compiling of various sources and contributors. More precisely, these recipe books were compiled from three different types of source:

1. content produced by copying and compiling of other written sources;
2. practical information obtained from personalities (practitioner or not) cited by the scribes;
3. content possibly derived from personal contributions made by the scribes.

In some instances, most of the content came from the copying and compilation of other written sources. This process can be followed by tracing the repeated appearances of certain popular texts found in the manuscripts of that period. Taking a wider view, these books have a great number of texts in common—dedicated to medicine, pharmacology, herbal, cosmetic, etc.—which were widely copied and disseminated in mediaeval and premodern times. These texts are quite often associated with the name of older or quoted authorities. Within our corpus several alchemical treatises and recipes are attributed to (pseudo) Albertus Magnus (c.1190–1280), Arnaldus de Villa Nova (c.1240–1311) or Roger Bacon (1214–1294). Previous studies have established that, quite often, such writings correspond to apocryphal or pseudepigraphical works.[21] As most recipe books are compilations, it is possible that some anonymous texts were (sometimes involuntarily) assembled together under the name of an authority cited in another part of the manuscript and subsequently disseminated under that name. Generally, these citations acted as a testimony of authority; they legitimised the alchemical knowledge recorded in these books. No doubt, the typical attraction and reverence for ancient authorities on the one hand, and the opportunity to record a (presumably) non anonymous text on the other hand, favoured the dissemination of these writings.[22] The association with the name of an authority gave rise to a tradition of works which, due to the processes of copying and compilation, circulated under various titles and were sometimes attributed to diverse authorities.[23]

At this stage, it should be noted that there are also a significant number of texts dedicated to religious content bound together with the recipe books under scrutiny. These are theological works, liturgies, extracts from the bible and hagiographies. In fact, a great number of recipe books appear to have been written or compiled within

[21] Minnis, *Theory of Authorship*. Concerning the alchemical works attributed to Albertus Magnus see notably Kibre, "Alchemical Writings." See also Newman, "Alchemy of Roger Bacon." For Arnaldus de Villa Nova, see notably Calvet, "Tradition alchimique latine."

[22] Minnis, *Theory of Authorship*, 9.

[23] Calvet, "Tradition alchimique latine," 42.

religious institutions, as attested by the citations of ownership. Signatures or monograms within these compilations indicate that these books were copied by scribes and members of this community. Obviously, the religious institutions—and their libraries—were privileged places, offering scribes the opportunity to copy and compile this kind of collection. The Munich Bayerische Staatsbibliothek Cgm 821, Cgm 822 and Clm 20174, formerly preserved in the Tegernsee monastery library, are good examples: they present not only similarities in terms of the different writings they contain but also, thanks to palaeographical analysis undertaken in the present study, it has been confirmed that several parts of their respective texts were recorded by the same scribe. This would imply that these manuscripts were (at least partially) copied in the same scriptorium, from similar written sources and by the same 'hand'.

Religious institutions may also appear as a contextual factor explaining the rapprochement of the various disciplines embedded within the manuscripts. Indeed, in general, medical and pharmaceutical recipes had an important place within religious communities. In this regard, art–technological recipes also found their place and could be linked with the art of writing and illuminating involved in scriptorial activity. The tables of contents of recipe books can be quite edifying on this point. For example, the table of contents in Munich Bayerische Staatsbibliothek Clm 20174 informs us that the artistic instructions were intended for the use of the scribes and illuminators of the scriptorium (*Et alia multa utilia per scriptoribus et illuministarum*, Clm 20174, fol. 1). In this context, scribe and illuminator, when not represented by the same person, worked side by side to produce manuscripts.[24] This collaboration led to enhanced communication and the development of a mutual interest in artistic practices among the monastic community.[25]

Practical or concrete interest and use of alchemical recipes in religious institutions is less obvious. It has been stated that writers of religious literature sometimes drew parallels with alchemical theories and processes.[26] Such writings, which obviously borrow alchemical vocabulary and imagery, are not included within our corpus. None of the alchemical texts under scrutiny were found to contain obvious religious connotations. But religious scribes' personal interest in alchemical craft procedures and practical alchemy in general can be attested by the large number of manuscripts produced that comprise both alchemical treatises and recipes. The presence of such instructions is more probably related to a certain attraction of alchemy for some monks or friars. Previous studies indeed have established that, even if the practice of alchemy was forbidden by several monastic orders, many of their members were at the root of alchemical (compilations of) texts and *Practica*.[27] Inventories of their library also inform us that they possessed alchemical treatises

[24] Cézard, "Alchimie et les recettes techniques," 6.

[25] Eamon, *Secrets of Nature*, 36.

[26] Principe & Newman, "Historiography of Alchemy," 398–400.

[27] Theisen, "Attraction of Alchemy." See also Barthélemy, *Alchimie de Guillaume Sedacer*, 26–8.

and recipe books.[28] Within our corpus, a relevant example is that of Wolfgang Seidel (1491–1562), prior but also copyist at Tegernsee monastery, who notably wrote two *Kunstbücher* (Munich Bayerische Staatsbibliothek, Cgm 4117 and Cgm 4118) between 1540 and 1550.[29] Cgm 4117 and 4118 reflect Seidel's interests in mathematics, astronomy, natural sciences and alchemy—disciplines in which he acquired theoretical but also practical knowledge. To do so, Seidel is known to have notably collected data from the libraries of Tegernsee but also from the neighbouring cloisters. During his stay at St. Ulrich's and St. Afra's Abbey (Augsburg), he made use of the abbey's vast collection of books, as attested in his commentaries recorded in Cgm 4118: "So many presents I have let copy from the library of the Cloister St Ulrich in Augsbourg, by a young boy whose name is Walthasar Gech von Fiessen in the year 1550."[30]

Seidel also seems to have relied on exchanges that are known to have taken place with contemporaries. In fact, in his *Kunstbücher*, he cites the authorities from whom he obtained practical information. These were either practitioners—artists—or contemporary scholars. For example, Seidel specifies several times that he is indebted to Bishop Philipp von Freising (1480–1541) for some recipes that he subsequently included in Cgm 4117. These prescriptions are notably dedicated to the melting of gold, silver and lead (Cgm 4117, fol. 2v, 37r–38v). Seidel also mentions Bartholome Schobinger (1500–1585), a jurist from St. Gallen.[31] The instructions recorded after Schobinger's name delineated a number of alchemical methods that notably serve to modify the properties of gold, to obtain a golden colour, and to work with gold, silver, iron and copper. Others concern the gilding on glass, the melting of ivory, metals and glass, the preparation of *aqua fortis* and the manufacture of a blue pigment called *azure* (Cgm 4117, fol. 62r–130r?).

These persons were learned persons or scholars, who were interested in natural philosophy and alchemy and who perhaps conducted their own experiments, as suggested by formulae which follow some of the recipes, such as *probatum vom Bischoff von Freising* (Cgm 4117, fol. 2v). Schobinger is notably at the root of a large compilation of alchemical texts.[32] He is also renewed for having personally known Paracelsus, who referred to Schobinger's writings.[33] The value of such an authority may appear visually in the recipe book. In the Cgm 4117, Seidel dedicates a whole page to recording Schobinger's name.[34] Moreover, the simple invocation

---

[28] See, for example, Barthélemy, "Alchimie et médecine," 110–3.

[29] Paulus, "Wolfgang Seidel"; and Pöhlein, *Wolfgang Seidel*.

[30] "So vill vom geschenckh hab ich auss der liberej des closters zw sant vlrich zw Augspurg lassen abschreiben durch ain knaben des namen ist Walthasar Gech von Fiessen im 1550 Jahr." (Munich, Bayerische Staatsbibliothek, Cgm 4118, fol. 128r).

[31] Schobinger, *Schowinger von St. Gallen*.

[32] *Allgemeine Deutsche Biographie*, 209; and Hertenstein, *Joachim von Watt*, 91–2.

[33] Meier, *Paracelsus*, 33–46.

[34] "Von bartholome Schobinger burger zu sanndt Gallen in Schweitz. Hab ich dise nachuolgende kunstel. etc./Empfangen den Sibenvnndzwaintzigisten tag. des Monats Februarii/Anno etc. 40." (Munich, Bayerische Staatsbibliothek, Cgm 4117, fol. 62r).

of the name of the Bishop of Freising would have served to confirm the efficacy of some of the technical instructions. Thus, the same way the scribes used to relate old treatises or data with the name of previous and quoted authorities, such as (pseudo) Albertus Magnus or Arnaldus de Villa Nova, they also mention those of their contemporaries to lend authority to validate the practicability or the reproducibility of the instructions they consign.[35]

In some cases, the information recorded in recipe books is documented as having been provided by an artist or practitioner. Augsbourg Staats- und Stadtbibliothek 2° Cod. 207 was produced in St. Ulrich and St. Afra's Cloister. It contains miscellaneous alchemical treatises and collections of recipes contributed by several scribes, including the monk Bild Vitus (1481–1529) and Johannes Gossolt (1421–1506), identified as *vicarius augustensis*.[36] In this work, Gossolt combined alchemical treatises attributed to (pseudo) Albertus Magnus with Latin and German alchemical recipes. For the latter he sometimes specifies his local sources. For example, at folio 171v, he mentions the "Magistri Jodoci Aurifabri de Haidelberga." Other citations of goldsmiths' names are found in our corpus of texts. In the St. Gallen Cod. Vadiana 395, several alchemical instructions are associated with the name of "Nicolaus Aurifaber." In many respects, metalworkers seem to have shared interest and knowledge in alchemical practices and materials.[37]

The scribes did not indicate how these data were actually provided and disseminated. At this stage, it is difficult to determine if these recipes were transmitted orally or only in written form. Oral transmission is usually favoured in specific contexts and environments in which people 'physically' converse.[38] In this regard, the workshop or laboratory probably offered the required closeness and the opportunity for oral exchanges and teaching. In the framework of this study, in only a few cases has it been possible to establish that a scribe *personally* met the authority he cited, meaning he might have obtained orally the practical information he recorded within his recipe book. This is notably the case for Seidel and two of the persons he cites, von Freising and Schobinger.[39] Nevertheless, it is quite unlikely that oral data circulated under the rhetoric of the recipe. This standardized and conventional textual format goes hand in hand with the copying process, and, thus, with a written transmission of knowledge. In other cases, exchanges in the form of correspondence are documented. For example, Seidel is also known to have exchanged letters with the monk Vitus, previously quoted, and (partially) responsible for Augsbourg Staats- und Stadtbibliothek 2° Cod. 207.[40] Both shared the same interest in natural philosophy, astronomy and alchemy—the same fields addressed within their writings.

[35] See notably Halleux, "Pratique de laboratoire."

[36] This hand is identified within the Augsbourg, Staats– und Stadtbibliothek, 2° Cod. 183, fol. 1r.

[37] Smith, *Body of the Artisan*, 140–51.

[38] Fox & Woolf, *Spoken Word*, 259–61.

[39] Pfaff, *Codex Vadiana*, 43.

[40] *Neue Deutsche Biographie*, vol. II, 235.

Finally, some recipes recorded within the corpus are a scribe's personal contribution. The acquisition of theoretical but also practical knowledge in natural science and alchemy may have lead Seidel to conduct his own experiments, which he then recorded in the form of recipes in his books. This possibility is confirmed in the first folio of Cgm 4118, where Seidel explains that he as well as both written (and older) sources and information collected from contemporaries, he had also drawn on his own practical experience.[41] The St. Gallen Ms. Vadiana 429 is an alchemical collection compiled between 1464/65 by Ulrich Ellenbog (1435–1499), a city physician in Ravensburg. A small part of its content also includes art technological recipes. Ellenbog's interest and practical knowledge in (al)chemy could notably be put in relation with his 1473 pamphlet *Von den giftigen besen Temppfen Reuchen der Metal* (On the poisonous and noxious vapours and fumes of metals). In this writing, the physician gives advices to goldsmiths and other metalworkers on how to protect themselves from the noxious effects of vapours of silver, mercury and lead.[42]

The Modalities of Composition

The diversity of sources and persons who contributed to these collections of recipes is evidenced by their varying modalities of composition. Codicological examination undertaken during this study has uncovered the (sometimes) very complex processes involved in the creation of recipe books.

A small number of these writings are produced in the form of carefully presented and independent collections: they are written in metallogallic ink and are quite often embellished with titles in red, and rubrics. These examples may be relatively homogeneous: usually, only one or two scribes (who are contemporaneous) can be identified and the presentation of their texts is almost identical. Moreover, no additional material modifies the original volume.

Others (though not the majority) are quite heterogeneous, both in their content (medical, theological, astronomical, technical, household) and in their physical appearance (diversity of format, dialect and handwriting). They are informally written, with no decoration, and are characterized by apparently random presentation and inconsistent structure.

The recipe titles, which do not always correspond to the procedure that follows them, do not imply a coherent organization. This second type of manuscript was compiled from several contributions and additions from various scribes and

[41] "De arte fusoria Rhapsodia partim ex uetusta quadam Biblioteca, partim uero bonorum amicorum colatione cum sumata, opera autem et labore fratris Wolffgangi Sedelij in vnum collecta in solacium et commodum fusorie artis studiosorum." (Munich, Bayerische Staatsbibliothek, Cgm 4118, fol. 1r).

[42] Teleky, *History of Factory*, 7; and Koelsch, *Geschichte des Arbeitsmedezin*, 101.

compilers, but also from the accumulation of physically distinct materials—quires and folios. Moreover, the diverse sections that make up these books often come from different geographical locations.

Frequently, additions and marginal notes attributed to the same scribe or to a later owner punctuate the distinct recipes within the manuscripts. In fact, these additions mostly appear under the form of titles or details given as a counterpart to the instructions. In the Ms. Vadiana 429 from St. Gallen, a great number of additional notes consist of technical commentaries and supplementary notes added to those of the compiler of the manuscript. In the Nuremberg Hs. 33733, a later owner added several titles and remarks within the margins. Some of these marginal additions also mention the name of the person from whom the scribe may have obtained the data he is adding. For example, in the Vienna Ms. 5224, fol. 74, the scribe indicated the name of a physician, "Doctor Jorg erffordie," before the title of a recipe dedicated to the production of sal ammoniac. On folio 105 of the same manuscript, the scribe associated an alchemical procedure with the name "Marggrauff von Rötell" by mentioning him in the upper margin. This observation provides a possible explanation for the considerable number of *unica* (isolated recipes) that appear only in one recipe book, and are thus likely to constitute data transmitted personally (and orally?) to the scribe.

The method of composition in this kind of recipe book indicates that they were compiled over a more or less long period, during or after peregrinations undertaken by their scribes. This is evidenced by notations mentioning different chronological periods and geographical provenances throughout the manuscripts. For example, Ms. 9715 from Nuremberg contains diverse collections of alchemical recipes. This manuscript was written by several scribes, who give names of persons or *magistri* underneath the practices they described. They also cite the different places where they collected their data and specify the dates of these events, which span several years. Notably there are several mentions of the "magistri Johannis Bog" and places such as "Erffordie" (Erfurt), and "Köln" (Cologne).[43]

Moreover, later additions or annotations found within the manuscripts tend to suggest that these books have been handled, manipulated and passed from one owner to another, sometimes over a long period. The *Prager Malerbuch* had several owners and circulated through several localities before entering the monastery of Zlatá Koruna. According to a note written by Federl Mir, the main scribe of the *Prager Malerbuch*, this manuscript was written c.1452, in Tittmoning in the district of Traunstein (Bavaria). This place probably corresponds to the original provenance of the recipe book. Moreover, the scribe tells us that he has gathered data from Michel Schril, a professor in Vienna, who passed away in 1472. We also know that from 1529 to at least 1599, this recipe book belonged to the Preisinger family. This family lived in Zettwing, in the present-day Czech Republic, between Munich and

[43] On Johannis Bog, see fol. 42v, 72v, 157v; on Erfurt, see fol. 49r; and on Cologne, see fol. 50v.

Vienna. Later, the manuscript is recorded within the inventory of the Zlatá Koruna convent, as indicated in folio 1r, where we find the date 1649.

Thus, the recording and disseminating of these instructions could go hand in hand with the circulation and penetration of alchemical and artistic knowledge outside the workshop or the laboratory.[44] It could be linked with a (partially oral?) transmission of knowledge that seems to have taken place between (learned) scribes, artists or artisans and scholars. Allusions to such exchanges are notably to be found in Seidel's *Kunstbücher*. For example, in Cgm 4117, fol. 1v, a recipe is stated as coming from a certain Thomas, caster in Munich, and transmitted via Freising to Seidel.[45] This instruction was placed in an available blank space, situated between the title of one of the book's sections and the table of contents (Fig. 1). It is credited to Seidel, but the handwriting is slightly different from the rest of the manuscript text. These observations suggest that this recipe, coming from a contemporary—perhaps oral—source is an isolated and later addition. Moreover, scribes sometimes even relate how contemporaneous authorities delivered their 'secret(s)' and even divulge the price they had to pay to obtain it. In other cases, recipes are recorded as being offered as a gift *pro memoria*.[46]

Contextualising the production and reception of these recipe books thus serves to highlight a large range of individual's personal's interest in alchemical and artisanal, as well as other types of knowledge. In this regard, a number of the recipe books produced in a religious institution are documented as having been later kept in a religious context. For example, Berlin Staatsbibliothek Theol. Lat. Quart. 152, written by "Frater Nicolaus lector" between 1408 and 1412, was owned by "Frater Polonus lector principalis" (Johannes Polonus), lector in the Thorn cloister during the fifteenth century. These manuscripts were usually moved to libraries at the beginning of the nineteenth century, during the period of secularisation that followed the French Revolution. In parallel, several examples of our corpus are documented as being part of private collections and were probably executed for or commissioned by a patron. This is notably the case for the *Kodex Berleburg* (Bad Berleburg, Schlossbibliothek Sayn–Wittgenstein, RT 2/6) which is recorded as being compiled for Bernhard of Breidenbach (c.1440–1497), who worked for the chapter of the cathedral of Mayence. Cod. Helm. 627 from Wolfenbüttel is a collection of alchemical treatises and instructions—including colour recipes—written around 1441–1444 by several hands. A note on the binding informs us that this volume probably belonged to the Bavarian physician Johannes Hartlieb (1410–1468), who wrote several compendia notably the *Puch aller verpoten kunst, ungelaubens und der zaubrey* (1456).[47]

---

[44] Halleux, "Alchimie," 342.

[45] "Vom Jungen thoman giesser zw munchen durch den bischoff von freising." (Munich, Bayerische Staatsbibliothek, Cgm 4117, fol. 1v).

[46] See Corbett, "Alchimiste Léonard de Mauperg."

[47] "Sum magistri Iohannis Hartliep, alias Walsporn, Vangionensis"; on Hartlieb, see Fürbeth, *Johannes Hartlieb*; for the edition of the text, see Eisermann & Graf, *Johannes Hartlieb*.

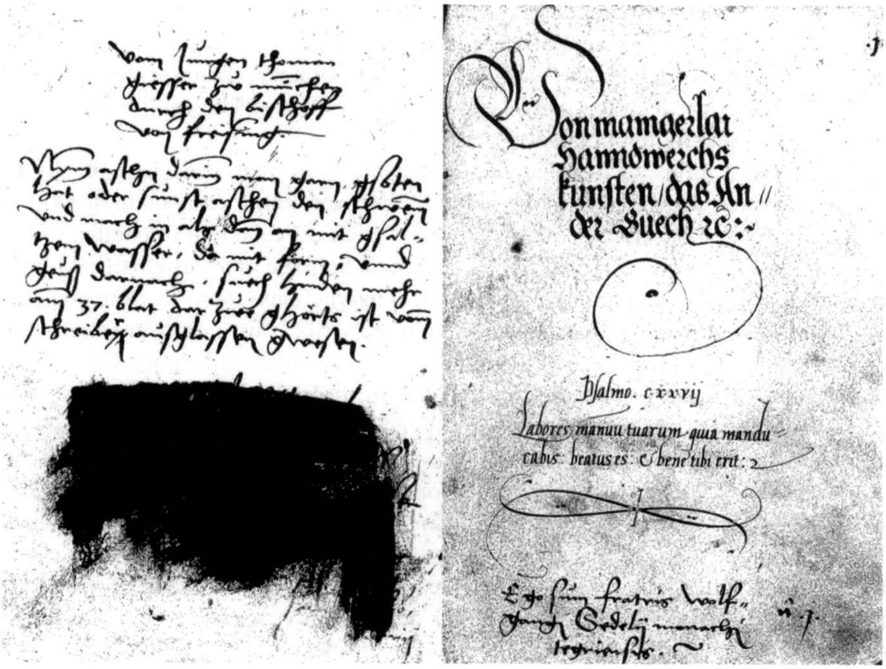

Fig. 1 Additional instruction due to Seidel, Munchen, Cgm 4117, fol. 1rv (Courtesy of Münchener Digitalierungszentrum)

The Function(s) of Recipe Books

The complex modalities of composition and diffusion of these texts raises some questions regarding their nature and their original function. At this stage, two different hypotheses have been put forward regarding the aim of this type of literature. On the one hand, these texts have been seen as manuals that may have been used by practitioners. On the other hand, the recipes often seem to have been transmitted for the purposes of literary preservation, not directly connected with contemporary workshop or laboratory practices.[48]

First of all, the textual environment and the diversity of the subjects bound together with the artistic and alchemical recipes in a same book, lead to the conclusion that these compilations were mainly read by scholars primarily interested in natural philosophy and were not intended for contemporary practical use. Moreover, it has been frequently stated that craft practices were transmitted orally,

[48] Clarke, "Codicological Indicators"; and Neven, *Recettes artistiques*, 16–23.

from the master to the apprentice.[49] A large number of the manuscripts of this study result from copying and compilation processes undertaken by scribes. As they were copied in a context outside the workshop or the laboratory, these recipes were not revised and, consequently, conveyed an anachronistic technical tradition that became more and more outdated.

Such observations seem to argue against the view that sees these books as manuals written for the practitioner. But neither were these compendium written purely for scholarly purposes, deprived of any practical function. In parallel to the data that could be considered part of the technical heritage of a earlier period, these recipe books also contain more recent practical instructions—coming from contemporary artists and practicing scholars or from the scribe's own experiments, as the examples of Seidel, Freising and Schobinger discussed above illustrate. Even when the writing of these instructions, verbalized in the rhetoric of the recipes, was carried out by scribes, data were not blindly copied. Scribes organised, assembled, completed or corrected when they felt it necessary. Thus, even if they were not the author *per se*, in the sense that they were not the origin or the source of the technical or chemical procedures they wrote down, they accomplished a set of activities linked to 'authorship'.[50] Scribes also made attempts to ensure that the recipes could be consulted at need: they composed tables of contents or indexes, they introduced titles within the margins and many other details which attest to a real desire to deliver usable information. In this context the marginal notes and additions made by the scribes/authors of the recipe book are of interest as most of them are technical comments testifying practical interest in both alchemical and artistic instructions. Several marginal annotations due to Seidel's hand punctuate the Cgm 4117 and consist in personal commentaries regarding the technical procedures he records. For example, on folio 53r, Seidel compares two ways for the melting of crystal. Concerning the first process he states in the margin that this 'art' was not of use to him as a better (method) is delivered on 219.[51] Then on folio 219, he indicates another method for the same technical procedure, giving as title "How one should masterfully melt crystal."[52]

In this sense, the scribes at the root of these recipe books created not simply a copy but a unique work, which reflected their own interests, their cultural and life context and sometimes their intention, which was to deliver practical and useful instruction.

[49] Halleux, *Entre technologie et alchimie*, 7.

[50] For this definition of authorship, see notably Love, *Attributing Authorship*, 32–40.

[51] "Dise kunst prauchet ich nit hinden amm 219 hastu vil pessere." (Munich, Bayerische Staatsbibliothek, Cgm 4117, fol. 53r)

[52] "Wie man christallen maisterlich giessen soll." (Munich, Bayerische Staatsbibliothek, Cgm 4117, fol. 219)

Reliability of Recipe Books

The modalities of composition and diffusion of these recipe books have an impact on their current (practical) use. During the compiling and disseminating processes, both alchemical and art-technological collections of recipes were subject to mutations, in the form of interpolation, reduction, contamination or assimilation with other texts. As the recipe books evolved and were modified by adding new texts and procedures, the recipes themselves could be modified in their technical formulations during their transmission from one manuscript to another. Assimilation with other texts occurs quite frequently, as the ingredients (and the actions) specified in these texts appear in the artistic recipe books but also in medical treatises, cookery books, and in alchemical or magical texts. Frequently, the copyist was free to add, to remove or to omit some words or even some parts of the text. These modifications or omissions sometimes concern primary data, such as the name of the ingredients or materials, or may be related to some of the steps of the procedure. At each stage of the copying process, variations or errors can occur. This phenomenon can be explained in several ways: it could be an attempt to improve or to diversify a previous formula; it could be a *quid pro quo*, in which an unknown or expensive ingredient is substituted with a more well known or less expensive one; it may have been a voluntary reduction of the recipe text.

If we suppose that the function of a recipe book was practical or instructive, this function could be the motivation behind changes to the recipes. An author or a scribe may, voluntarily, have corrected the text, or added information to it. However, changes to the recipe may also be due to a misunderstanding of the procedure. Such miscomprehension may be due to palaeographical problems that resulted in a word being misread or misunderstood and thus replaced by another. This was a likely occurrence if the copyist was not a practitioner or if he was not able to translate or to decipher an unreadable formula. For example, in Heidelberg Cod. Pal. Germ. 183, fol. 286, at the beginning of a recipe dedicated to the production of minium, the scribe mentions the use of "*Lautterm sapienticum*" instead of *Lutum sapientium.* In Munich Bayerische Staatsbibliothek Cgm 824, the scribe describes the preparation of a white (fol. 13r), a yellow (fol. 13r), a blue (fol. 14v) and a grey pigment (fol. 14v), and each time suggests taking "cretam *rosam.*"[53] The same instructions are recorded in the Cgm 822 (fol. 64v) where the scribe correctly indicates the use of *cretam rasam* (scraped chalk).

Such phenomena—reduction, amplification, variation—may result in a procedure whose description can seem vague or unclear and thus thwart the current use and relevance of recipe books in the study and the reconstruction of historical artistic practices.

[53] My italics.

Alchemical and Art-Technological Recipes Within a Manuscript: Location, Relationship and Distinction

Similarities of format and modalities of composition and diffusion may have had an impact on the recording and assembling of alchemical and art-technological recipes within the same manuscript. This could notably result in the mixing and grouping of different types of unrelated instructions.

More precisely, in the corpus under scrutiny, alchemical instructions appear either as independent pieces of work, or as isolated (groups of) recipe(s) embedded with artistic or other types of instructions. In the first case, alchemical content may appear concurrently with an artist's recipe book within the same manuscript but in a separate section. When this occurs, the texts mostly consist of quite theoretical alchemical treatises, often associated with the name of a former or contemporary authority. Most of them are attributed to the (pseudo) Albertus Magnus, Roger Bacon and Arnaldus de Villa Nova whose writings date from an earlier period. These works could also be 'physically' distinct works, delimited to a quire or a booklet—or even a folio—and assembled with the rest of the manuscript at a contemporary or later period. Vienna Ms. 5224 contains various alchemical collections of recipes and *practica,* all of which are delineated and separated by blank spaces or folios. These texts were written by several hands, on paper from different origins dated from the fifteenth and the sixteenth centuries. The main contribution comes from an anonymous scribe who is responsible for a number of independent collections of recipes but also for some additions throughout other parts of the volume.[54] Perhaps this contributor was at the root of both the (partial) writing and the collecting and assembling of these data into one single volume. This theory is supported by the fact that his hand dates from the sixteenth century, which coincides with the estimated date of the binding and the titles written on the cover. Once all the diverse parts were bound together, the manuscript was subject to later additions by the main scribe, who wrote these on previously blank space (fol. 143v–144r and fol. 158r, 163v), both at the beginning and the end of two distinct treatises.

Alchemical texts are also sometimes situated alongside an artists' recipe book, either before or after. If this is the case, they will be found next to technical instructions dedicated to procedures similar to those described in an alchemical context, such as the imitation of gold or silver, the gilding of stones or glass, the manufacture of vermilion, the purification of ultramarine, the melting of stones or metals, or several dyeing procedures. The alchemical content may be delimited within the title(s), chapter(s) or table of contents or 'physically' circumscribed by a folio or a quire. But, in most cases, there is no obvious delimitation between the two distinct collections of recipes. For example, in Nuremberg Germanisches

[54] Identified as 'hand' 4 in the catalogue notes, he is responsible for fol. 31v, 38r–120v, 123r–143r, 153r–157v, 160r–163r.

Nationalmuseum Ms. 5078b (fol. 2r–41r), the scribe moves from one subject to another with no indication that the subject has changed. Moreover, this set of alchemical and art-technological recipes is followed with no clear distinction (no title, nor blank space) by a series of medical prescriptions due to the same hand.

In another case—isolated (groups of) alchemical recipes found in the middle of prescriptions of another type—detection of alchemical content and distinction from artistic instructions within recipe books can be fraught with difficulty. Similarities in terms of their textual format probably lead scribes to group them with other sort of prescriptions. When found as isolated elements, alchemical and art technological recipes usually appear within a large broad of various (and unrelated) writings. For example, part of Nuremberg Hs. 3227 (fol. 74v–81v, 90v–164v) is a miscellanea of cooking, alchemical, household and artistic recipes, written by the same hand. Heidelberg Cod. Pal. Germ 678 notably includes a collection of medical recipes interrupted by one single alchemical recipe, dedicated to the manufacture of vermilion. In Berlin Theol. Lat. Quart. 152, some isolated alchemical recipes are placed in the middle of several cooking recipes and within religious texts.

Finally, some recipes were never granted their own place within a collection of recipes. An isolated recipe is sometimes jotted down on any available space on a page or squeezed into an even less appropriate place. For example, in Nuremberg Ms. 27773, recipes dedicated to the colouring of glass and the hardening of steel appear under the form of later additions in the upper and lower margin of a school book, and probably also on the binding board.

Thus reading these collections and attempting to categorise the recipes as alchemical or art-technological can be less than straightforward. After examining the corpus in question the following suggestions are proposed to help identify the different recipes.

As stated above, whether alchemical or art-technological, the recipes contained in these manuscripts are presented in the form of a *formula* which, in most cases, enumerates the ingredients and the actions necessary to produce a particular preparation. In addition artistic recipes sometimes indicate the recommended geographical provenance or grade of quality of the ingredients. Suggestions for possible substitutions might also appear. This sort of information is rare in alchemical recipes.

The length of a recipe depends not only on the number of ingredients involved but also on its complexity, the number of steps necessary to obtain the final product. A recipe can be anything from one sentence to several pages within a manuscript. Alternatively, a recipe may appear merely as a brief list of ingredients, without any other additional information. In fact, two categories of recipe can be distinguished: the *Vollrezepte* (detailed recipes) and the *Kurzrezepte* (abbreviated recipes).[55] In the first, the quantities and the various steps are indicated. In the second, only the

[55] Halleux, "Alchimie," 343.

ingredients are cited: the procedure is sketched out or omitted altogether and the rest is left to the ingenuity of the user. This second category is more common in the case of artistic recipes; a great many of the recipes dedicated to the manufacture of ink are written in the form of a very short list of ingredients. It is less common for the alchemical recipes to be presented this way.

The title of a recipe may also give an indication of the final product to be obtained and, in some cases, specify the use of the product. Again, this is particularly true for artistic instructions and is less observable for alchemical ones.

For both types of instructions (alchemical or artistic), some steps could be omitted or were left to the interpretation of the reader. Specified quantities may be missing in both fields. When quantities are given, artistic recipes are far more likely to use local measurements, whereas in alchemical instructions, the quantities—if mentioned at all—are more often expressed in terms of ratio or proportions. In some cases, these proportions are not 'practically' correct. A very well-known example is the proportion of mercury and sulphur proposed by mediaeval recipes for the production of vermilion which is invariably incorrect.[56] Very rarely are the correct chemical proportions cited.

In parallel, alchemical writings may involve the use of symbols or metaphors to designate substances and practices. In consequence, the way an alchemical recipe was received would depend on the degree of experience of the reader-practitioner reading it. On the one hand, the (sometimes) metaphorical or codified language as well as the approximations stressed the arcane nature of these recipes and contributed to their secrecy. On the other hand, the omitted information may have been complemented by data only known to some readers and not recorded by the copyist who conserves only the essential part of the recipe. If so then this kind of recipe was only meant to be accessible and useable by those practitioners who could easily fill in the lacuna that punctuated the text of the recipe.

As previously observed, citations of authority were frequently used by the scribes of the manuscripts. However, the tendency for an older authority to be cited in the recipe books is particularly characteristic of the alchemical writings and less typical of the art-technological recipes. As stated above, such citations primarily served to legitimate the technical and chemical procedures. In addition, most alchemical recipes describe processes and practical results to validate previously enounced theoretical principles. Thus, more than artistic recipes, alchemical instructions emphasize the efficacy of the procedure which is frequently confirmed through expressions such as *expertum es* or *probatum est* which are placed at the beginning or end of the instructions. The notion of *experimenta* (testing) implies the acquisition or confirmation of theoretical knowledge through direct observation and experimentation rather than through analysis based on rational arguments.[57] In such a case, when one of these reassuring expressions appears at the end (or the beginning) of a recipe, it does not signify that the recipe has actually been tested by the scribe.

56 Bucklow, "Paradigms and Pigment Recipes," 144.

57 Halleux, "Pratique de laboratoire."

Rather it implies that the recipe constitutes a plausible set of instructions, and has been successfully performed at least once and/or confirmed by a previous authority.

Particular interest in the empirical aspects of the technical procedures is also more perceptible within alchemical instructions, in comparison to artistic recipes. The former pay greater attention to the chemical aspects of the craft process they detail and describe more precisely each stage of the transformation of matter, from original to final (and more perfect) form. Whereas artists such as painters were interested in the physical appearance of their materials, alchemists were more interested in the fundamental changes that might occur within the matter. This could perhaps be related to Paul of Taranto's description in the *Theorica et practica* of primary and secondary qualities and the distinct ways artists and alchemists worked on substances. He considered artists capable of producing only 'extrinsic' or external changes as they operate on secondary or 'artificial' qualities such as colors. In contrast, alchemical attempts to manipulate primary qualities transmute substances intrinsically and operate fundamental change.[58]

Accordingly, alchemical recipes also dedicate a large part of their text to the description of chemical apparatus, tools and containers. These writings pay particular attention to the use of a variety of containers and receptacles and their specific purposes. Moreover, in several manuscripts, such as Seidel, Vienna Ms. 5224 or Cod. Helm. 627 from Wolfenbüttel (to cite but a few), the text is punctuated by illustrations of these (Fig. 2). This is rare, if not non-existent in artistic instructions.

These last distinguishing features of alchemical, as opposed to art-technological texts, go hand in hand with the fact that practical alchemy relies on theoretical (or speculative) principles. Quite often, these recipes should be seen and understood as *experientia* which are meant to serve as a rational demonstration of a preceding *theorica*.[59]

Conclusion

Examination of the processes of making, compiling and disseminating this corpus of mediaeval and premodern recipe books provides us with information concerning their nature and former function. One the one hand, it has been established that these manuscripts were mostly written in religious centres and that they are largely the result of the copying process undertaken by scribes. Moreover, within these books, alchemical and artistic recipes were frequently recorded alongside a wide range of various—and *a priori* unrelated—subjects which may have been written by the same person. Both the context of their production and the similarities in

58 Newman, "Technology and Alchemical Debate," 434, 442–5. The author largely relies on Paul of Taranto, *Theorica et practica*, Paris, BN, Lat. 7159, fol. 1r–55r for which he delivers a partial edition and translation.

59 Halleux, "Pratique de laboratoire," 118–22.

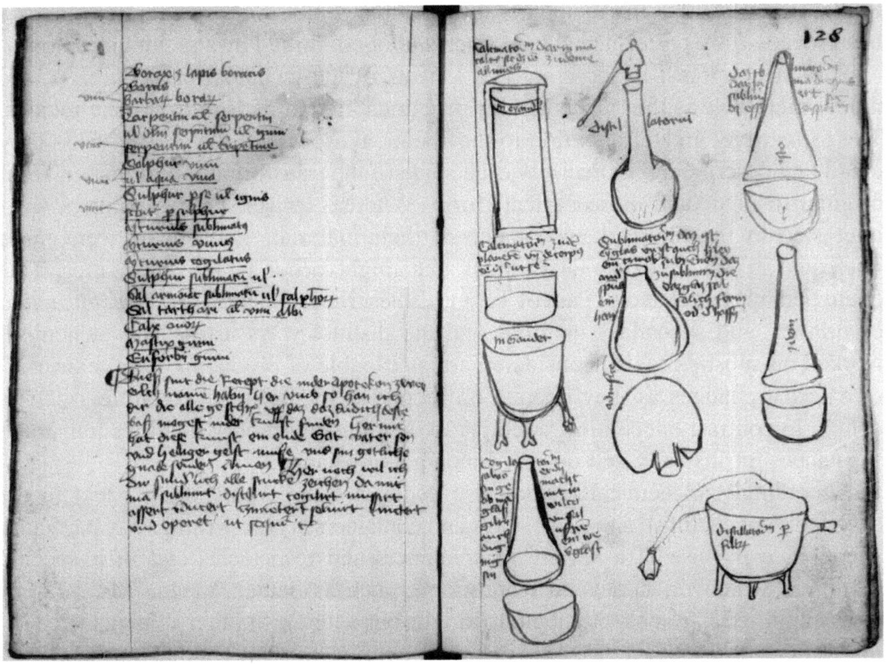

Fig. 2 Illustration of containers within an alchemical text, Wolfenbuttel, Cod. 627, fol. 127v–128r © Photographer (Courtesy of Wolfenbuttel Library)

terms of their textual format could serve to explain their propinquity. These first observations tend to suggest that these recipe books were produced for literary purposes and to preserve existing knowledge. And, indeed, these compilations were mainly read by a scholarly public primarily interested in natural philosophy, astrology, and alchemy and were probably not intended for practical use within the workshop or the laboratory.

Moreover, as these books are the result of compilation and additions of data, the finding and the delimitation of alchemical content can be complicated, especially when isolated (groups of) recipe(s) were recorded in the middle of unrelated (collection(s) of) text(s). By displaying the various ways alchemical and artistic recipes are embedded within the same manuscript, this study has highlighted the potential difficulties in localizing and distinguishing them.

On the other hand, it has been demonstrated that recipe books partly derived from the recording and transmission of (more or less) contemporaneous practices. Thus, recipe books also reflect alchemical and artistic knowledge and interests of both scribes and contemporary scholars, both of whom could be involved as readers or authorities. Recipe books also serve to define a more precise network in which these types of knowledge circulated, delivering information about the 'actors'—whether artisans, scholars, natural philosophers, (theoretical) alchemists or lay scribes—and their interconnections, as well as the media (copy, oral source, experiment) they used to exchange, share and communicate art and alchemy.

Appendix: List of Manuscripts

Augsbourg, Staatsbibliothek

– *2° Cod 207*, c.1514
Scribe: Johannes Gossolt and Bild Vitus (1481–1529), monk at St Ulrich in Augsburg
Language: Latin and German

– *2° Cod 572*, before 1446 (part 2)–1446 (part 1)
Language: partly written in Swabian (part 1) and Bavarian (part 2) dialects

– *4° Cod 131*, 15th–16th century (the recipes)
Language: German

– *4° Cod 149*, c.1501–1519
Scribe: Leonhard Wagner
owner:
Language: Schwabian
Origin: written in Augsbourg (St Ulrich and Afra), Irsee, St Gallen, Lorsch

Bad Berleburg, Schlossbibliothek Sayn-Wittgenstein

– *RT 2/6 Kodex Berleburg*, c.1475–1478
Language: Franconian and Latin
Origin: Rhine Main
Previous owner: Bernhard of Breidenbach, (who worked for the chapter of the Cathedral of Mayence)

Bamberg, Staatsbibliothek

– *L III 33*, 16th century
Language: Middle German

Berlin, Staatsbibliothek

– *Germ. Fol. 8*, c.1430–1440
Language: Swabian, Latin and Italian. The text is written in different hands including that of Johannes Seiler.
Origin: South of Germany, Switzerland or Bohemia

– *Germ. Quart. 15*, 1496 (fol. 156)
Language: Latin and German
Origin: South of Germany

– *Theol. Lat. 152*, 1408 and 1412
Origin: Torgau and Dresden (main text)
Scribe: 'Frater Nicolaus lector' (fol. 121r, 132r, 140v) in 1408 in Torgau and 1412 in Dresden.
After that, the ms. is documented as being in Thorn, the 5 of March 1427.
Previous owner: Johannes Polonus ('Frater Polonus lector principalis'), Lector in the Thorn cloister (15th century)

Budapest, Nationalbibliothek

– *Cod. Germ. 36*, 1487–1492
Language: Alemanic and Latin

Erfurt, Bibliothek der Stadt

– *Amplonius Quart. 189* ('Notae de coloribus Liber de coloribus et virtutibus lapidum, Pseudo-Albertus Magnus Lapidarium, De coloribus, naturalia exscripta et collecta'), 13th–14th century
Origin: Mainz (?) according to a mention associated with the date of 'December 1407'

Heidelberg, Universitätsbibliothek

– *Cod. Pal. Germ. 183*, 1560–1570/71
Scribe: Michel (?)
Language: High German including Bavarian features
Provenance: Amberg, preserved in the Amberger library of Ludwig VI, Count Palatine, according to inscription on the binding board: 'H[erzog] L [udwig VI.] P[falzgraf] 1570'

– *Cod. Pal. Germ. 678,* 15th century
Origin: South West Germany

– *Cod. Pal. Germ. 696,* ('Die kunst glaß zu schmeltzen und gießen von haugen von wildpürg simmerischer Amptmann'), 16th century

Karlsruhe, Badische Landesbibliothek

– *Cod. R 49*, 15th century, mention of 1465
Language: Swabian dialect

Munchen, Bayerische Staatsbibliothek

– *Cgm 821*, ('Liber illuministarius, pro fundamentis auri et coloribus ac consimilibus'), c.1500–1512 (for the second part)
Scribe: Konrad Sartori (scribe at Tegernsee Monastery)
Language: Latin and Bavarian
Origin: Tegernsee Monastery

– *Cgm* 822, 14th–with additions from 15th century
Language: Latin, Bohemian, Bavarian, middle German and Swabian dialects
Origin: mention of several Augsburger painters. Exlibris of the Tegernsee library 1485 (fol. 1v)

– *Clm 405*, c.1390 (addition in 15th century)
Language: Latin and Alemanic
Previous owner: Bishop Guido de Valencia (from Tripoli) according to fol. 1r. The manuscript was in Osthoven in 1461 (fol. 25 'Subscriptio filii Heinrici Aysinger in Osterhoven a. 1461')

– *Clm. 444*, ('Tractatus de coloribus faciendis. De cerusa componenda...Accipe laminas plumbeas vel stagneas'), 14th–15th century
Language: Latin

– *Clm. 7623*, 14th century (beginning)
Language: Latin and German

– *Clm. 20174*, 1464–1473
Language: Latin and German
Origin: Ex-libris of Tegernsee Monastery, 1482

Nuremberg, Germanische Nationalmuseum

– *3227a*, c.1389 (additions from 15th century)
Scribe/ author: partly written by 'Hanko pfaffen Doebringers' (according to a mention on fol. 43r)
Language: Latin, Bavarian and Middle German dialects
Provenance: Cologne/ mention of 'Nicolaus Pol doctor 1494'

– *5078b*, 15th century
Language: Middle Bavarian
Origin: Bavaria

– *9715*, 15th century
Origin: Bavaria

– *27773*, c.1260 (addition in mid-14th century)
Origin: Marbach—the manuscript was bound before 1354 in the canon order of St Augustin in Marbach

– *33733*, c.1455–1457
Language: Bavarian
Previous owner: fol. 1r '15R74 Siluester Schafman von Hamerberg I-B-G (?)'

– *141871*, 16th century (beginning)
Language: Middle German

– *147699*, c.1488–1490
Language: Swabian and Bavarian dialects

Prague, Narodni Knihovna

– *Cod. XI D 10*, c.1452–1477
Scribe: Federl Mir (1452)
Language: Bavarian and Latin
Origin: Tittmoning
Previous owner: Preisinger Family (1529–1599) from Zettwing, Sancta Corona monastery (1649)

St Gallen, Kantonsbibliothek

– *Vad. 395*, 15th and 16th century
Language: German and Latin

– *Vad. 407*, c.1522
The main scribe signed at fol. 155: 'Michel Cochemus 1522' and fol. 253v : 'Michael Cochemus 1522'.
Language: German

– *Vad. 429*, c.1465
Origin: South of Germany
Previous owner: Ulrich Ellenbog

Trier, Stadtbibliothek

– *1024/1936*, ('De coloribus et mixtionibus-Incipit libellus Mappe clauicula dictus'), 15th century, mention of 1437
Origin: Trier (?)

Vaticano, Biblioteca Apostolica Vaticana

– *Pal. Lat. 1330*, 1463–64
Scribe: Walpod, Heinrich (active for Nikolaus of Kues)
Language: Latin
Previous owner: Johannes of Bavaria, canon in Augsburg (1477)

Vienna, Österreichische Nationalbibliothek

– *5224*, 1481 with 16th century additions (fol. 31v, 38r-120v, 123r-143r, 153r-157v)
Language: Latin and German

– *5489*, 14th–15th century, mention of 1462 (fol. 180v), 1463 (fol. 146r) and 1464 (fol. 218v)
Language: Latin and Bavarian

– *5509*, 15th century, mention of 1459 and 1464
Language: Bavarian

Winterthur, Stadtbibliothek

– *Cod. 4° 47*, ('Hie vachet an ein bewerte edle kunst und nützliche wie man sol ferwen lini tuoch wullin tuoch faden garn mitt allen farwen die da gerecht sind und wie man sÿ zuo venedig ferbt'), 15th–16th century, mention of 1575 and 1579
Scribe: Haymhofer Thomas, from Basel
Language: German

Wolfenbuttel, Herzog- August Bibliothek

– *Helmst. 627*, 15th century, c.1444
Origin: mention of Heidelberg, 1444
Previous owner: belonged to the Bavarian physician Johannes Hartlieb (1410–1468) 'Sum magistri Iohannis Hartliep, alias Walsporn, Vangionensis'

Zürich, Stadtbibliothek

– *B 245*, 15th century
Language: Middle German

Bibliography

Allgemeine Deutsche Biographie. Leipzig: Duncker & Humblot, 1891.
Neue Deutsche Biographie, vol. 2. Otto Graf zu Stolberg-Wernigerode, 1955.
Barthélemy, Pascale. 2002. *La Sedacina ou l'œuvre au crible. L'alchimie de Guillaume Sedacer, carme catalan de la fin du XIVe siècle*. Paris/Milan: S.É.H.A-Archè.
Barthélemy, Pascale. 2003. Les Liens entre alchimie et médecine. L'exemple de Guillaume Sedacer. *Micrologus' library. Alchimia e medicina nel Medioevo* 9: 109–134.
Berthelot, Marcellin, and Charles E. Ruelle. 1888–1889. *Collection des anciens alchimistes grecs*. Paris: G. Steinheil.
Berthelot, Marcelin, and Rubens Duval. 1893. *Histoire des Sciences: La Chimie au Moyen-Âge*, vol. 2: *L'Alchimie syriaque*. Paris: Imprimerie nationale.
Brewer, John S. 1859. *Fr. Rogeri Bacon Opera quaedam hactenus inedita*. London: Longman.
Bucklow, Spike. 1999. Paradigms and Pigment Recipes: Vermilion, Synthetic Yellows and the Nature of Egg. *Zeitschrift für Kunsttechnologie und Konservierung* 13: 140–149.
Calvet, Antoine. 2007. La Tradition alchimique latine (XIIIe–XVe siècle) et le corpus alchimique du pseudo-arnaud de Villeneuve. *Médiévales. Le Livre de science, du copiste à l'imprimeur* 52: 39–54.
Cézard, Pierre. 1945. L'Alchimie et les recettes techniques. *Métaux et civilisations* 1: 5–10.
Clarke, Mark. 2001. *The Art of All colours. Medieval Recipe Books for Painters and Illuminators*. London: Archetype Publications.
Clarke, Mark. 2009. Codicological Indicators of Practical Medieval Artists' Recipes. In *Sources and Serendipity. Testimonies of Artists' Practice*, ed. Erma Hermens and Joyce Townsend, 8–17. London: Archetype Publications.
Corbett, James A. 1936. L'Alchimiste Léonard de Maurperg (XIVe siècle). Sa collection de recettes et ses voyages. *Bibliothèque de l'école des chartes* 97: 131–141.
Eamon, William. 1994. *Science and the Secrets of Nature. Books of Secrets in Medieval and Early Modern Culture*. Princeton: Princeton University Press.
Eis, Gerhard. 1962. *Mittelalterliche Fachliteratur*. Stuttgart: Metzler.

Eisermann, Falk, and Eckhard Graf. 1998. *Johannes Hartlieb: Das Buch der verbotenen Künste*. München: Diederichs.

Fox, Adam, and Daniel Woolf. 2002. *The Spoken Word. Oral Culture in Britain, 1500–1850*. Manchester: Manchester University Press.

Fürbeth, Frank. 1992. *Johannes Hartlieb, Untersuchungen zu Leben und Werk*. Tübingen: M. Niemeyer.

Haage, Bernhard D., and Wolfgang Wegner. 2007. *Deutsche Fachliteratur der Artes in Mittelalter und früher Neuzeit*. Berlin: Erich Schmidt.

Halleux, Robert. 1979. *Les Textes alchimiques*. Turnhout: Brepols.

Halleux, Robert. 1981. *Alchimistes grecs*, vol. 1: *Papyrus de Leyde. Papyrus de Stockholm. Fragments de recettes*. Paris: Les Belles Lettres.

Halleux, Robert. 1988a. L'Alchimie. *Grundriss der romanischen Literaturen des Mittelalters, 8/1, La littérature française aux XIVe et XVe siècle* 16: 336–345.

Halleux, Robert. 1988b. Pratique de laboratoire et expérience de pensée chez les alchimistes. In *Zwischen Wahn, Glaube und Wissenschaft: Magie, Astrologie, Alchemie und Wissenschaftsgeschichte*, ed. Jean-François Bergier, 115–126. Zürich: Verlag der Fachvereine.

Halleux, Robert. 1989. Entre technologie et alchimie. Couleurs, colles et vernis dans les anciens manuscrits de recettes. In *Technologie industrielle. Conservation, restauration du patrimoine culturel, colloque AFTPV-SFIIC*, Nice, 19–22 septembre 1989, 7–11. Paris: Erec.

Halleux, Robert. 2009. *Le Savoir de la main, savants et artisans dans l'Europe pré-industrielle*. Paris: Armand Colin.

Halleux, Robert, and Paul Meyvaert. 1987. Les Origines de la Mappae clavicula. *Archives d'Histoire Doctrinale et Littéraire du Moyen-Age* 54: 7–58.

Hedfors, Hjalmar. 1932. *Compositiones ad Tingenda Musiva, herausgegeben, übersetzt und philologisch erklärt*. Uppsala: Almqvist & Wiksell.

Hertenstein, Bernard. 1975. *Joachim von Watt (Vadianus), Bartholomaüs Schobinger, Melchior Goldast: die Beschäftigung mit dem Althochdeutschen von St Gallen in Humanismus und Frühbarok*. New York: De Gruyter.

Jansen-Sieben, Ria. 1989. *Repertorium van de middelnederlandse artes-literatuur*. Utrecht: Hes & De Graaf.

Johnson, Rozelle P. 1939. *The Compositiones Variae from Codex 490, Biblioteca Capitolare, Lucc*. Urbana: University of Illinois Press.

Kahn, Didier. 2007. *Alchimie et paracelsisme en France à la fin de la Renaissance (1567–1625)*. Genève: Droz.

Kibre, Pearl. 1942. Alchemical Writings Ascribed to Albertus Magnus. *Speculum* 17: 499–518.

Koelsch, Franz. 1967. *Beiträge zur Geschichte des Arbeitsmedizin*. Munich: Bayerische Landesärztekammer.

Laurioux, Bruno. 1997. *Les Livres de cuisine médiévaux*. Turnhout: Brepols.

Love, Harold. 2002. *Attributing Authorship: An Introduction*. Cambridge: Cambridge University Press.

Martelli, Matteo. 2011. *Pseudo-Democrito. Scritti alchemici con il commentario di Sinesio*. Paris/Milan: S.É.H.A.-Archè.

Meier, Pirmin. 2004. *Paracelsus. Arzt und Prophet*. Zürich: Ammann.

Minnis, Alastair. 2010. *Medieval Theory of Authorship. Scholastic Literary Attitudes in the Later Middle Ages*. Philadelphia: University of Pennsylvania Press.

Neven, Sylvie. 2011. *Les Recettes artistiques du Manuscrit de Strasbourg et leur tradition dans les réceptaires allemands des XVe et XVIe siècles (Étude historique, édition, traduction et commentaires technologiques)*. Liège: Université de Liège.

Newman, William R. 1989. Technology and Alchemical Debate in the Late Middle Ages. *Isis* 80: 423–445.

Newman, William R. 1994. The Alchemy of Roger Bacon and the Tres Epistolae Attributed to Him. In *Comprendre et maîtriser la nature au Moyen Âge. Mélanges d'histoire des sciences offerts à Guy Beaujouan*, ed. Danielle Jacquart, 461–479. Genève: Droz.

Paulus, Nikolaus. 1894. Der Benediktiner Wolfgang Seidel. Ein bayerische Gelehrter des 16. Jahrhunderts. *Historisch Politische Blätter* 113: 165–185.
Pereira, Michela. 1999. Alchemy and the Use of Vernacular Languages in the Late Middle Ages. *Speculum* 74: 336–356.
Pfaff, Silvia. 1994. *Der Codex Vadiana 404 von Wolfgang Seidel (1492–1562) – Erster Teil seines dreibändigen Kunstbuches?* Munich: Technische Universität München.
Pöhlein, Hubert. 1951. *Wolfgang Seidel, Benediktiner aus Tegernsee, Prediger zu München Sein Leben und sein Werk.* Munich: Zink.
Principe, Lawrence M. 2013. *The Secrets of Alchemy*. Chicago: University of Chicago Press.
Principe, Lawrence M., and Lloyd DeWitt. 2002. *Transmutations: Alchemy in Art*. Philadelphia: Chemical Heritage Foundation.
Principe, Lawrence M., and William R. Newman. 2001. Some Problems with the Historiography of Alchemy. In *Secrets of Nature: Astrology and Alchemy in Early Modern Europe*, ed. William R. Newman and Anthony Grafton, 385–431. Cambridge: MIT Press.
Schobinger, Viktor. 1978. *Die Schowinger von St. Gallen, Bartlome Schowingers Bücher.* Zürich: Schobinger.
Singer, Dorothea W. 1928–1931. *Catalogue of Latin and Vernacular Alchemical Manuscripts in Great Britain and Ireland*, 3 vols. Brussels: Lamertin.
Smith, Pamela H. 2004. *The Body of the Artisan: Art and Experience in the Scientific Revolution.* Chicago: University of Chicago Press.
Smith, Cyril S., and John G. Hawthorne. 1974. *Mappae Clavicula. A Little Key to the World of Medieval Technique*. Philadelphia: American Philosophical Society.
Teleky, Ludwig. 1948. *History of Factory and Mine Hygiène*. New York: Columbia University Press.
Theisen, Wilfrid. 1995. The Attraction of Alchemy for Monks and Friars in the 13th–14th centuries. *The American Benedictine Review* 46: 239–253.

Artisanal Processes and Epistemological Debate in the Works of Leonardo Da Vinci and Vannoccio Biringuccio

Andrea Bernardoni

Abstract During the Renaissance the field of the chemical arts was characterised by the complex identity of its protagonists, who would hardly recognize themselves in traditional socio-cultural and professional definitions. Although it is anachronistic to speak of chemistry as a discipline, during the late fifteenth and early sixteenth centuries it is possible to outline an area of interest around the transformation processes of substances in which various fields of theoretical and practical knowledge intersect. In this cultural context the figures of Leonardo da Vinci and Vannoccio Biringuccio stand out as, in addition to their experimental activity, they became the promoters of a general reform of chemical arts and natural philosophy.

'Chemical' Arts During the Renaissance

Talking about chemistry in relation to the Renaissance calls for historiographical considerations, as chemistry did not exist as an independent art or discipline during this period. In Latin and vernacular languages the term 'chemistry', was first used during the sixteenth century. In the *De sculptura* by Pomponio Gaurico (1482–1530), published in Florence in 1504, *chemiké* is used to refer to the preparation of moulds for the ancient art of casting and in the process of pouring alloy to make sculptures. Gaurico considers *chimice* a "dirty and smoky activity that uses clay, dung, charcoal and bellows."[1] Other authors used chemistry as a synonym of

[1] "Nunc vero de altera parte dicendum esset, quam nos *chemickén* sive fusoriam nominavimus. Quae quoniam et foeda et caliginosa est, eam, si videbitur, praetarmittamus. [. . .] Chimice unde infamis illa omnibus usitatissima saeculis ars dicta, quae circa metallorum exalterationem versatur. in duas a nobis dividetur partes: in formarum confectionem et in metallorum

A. Bernardoni (✉)
Museo Galileo - Institute and Museum for the History of Science, Florence, Italy
e-mail: a.bernardoni@museogalileo.it

S. Dupré (ed.), *Laboratories of Art*, Archimedes 37,
DOI 10.1007/978-3-319-05065-2_3,

alchemy.[2] In Georg Agricola's (1494–1555) books, for instance, *chymistae* and *chymia* mean 'alchemists' and 'alchemy' respectively.[3] In the chapter "Chymica" in Girolamo Cardano's (1501–1576) *De rerum veritate*, *chymistae* is used synonymously with 'alchemists' in relation to the manipulation of matter by the arts, particularly metals and gems.[4]

Taking an etymological perspective, *chymia* and *alchymia* have the same meaning because they both derive from *Chem* in ancient Egyptian, the former via Greek, and the latter via Arabic.[5] In this sense, from a terminological point of view the problem of anachronism for the use of the term 'chemistry' in the sixteenth century and its differentiation from 'alchemy' disappears, because *chymia* is the art of metal transformation derived from Greek etymology, *alchymia* is the same term from Arabian etymology. This is clearly specified by Tommaso Garzoni (1549–1589) in his *La piazza universale di tutte le professioni del mondo:* "[. . .] all those who have reasoned, or reason against alchemy, and who assume 'alchymia' from Arabian or 'Chimia' from Greek [. . .]."[6] Vannoccio Biringuccio (1480–c.1537), as we will see later, defined a generic domain for the arts of fire, considering alchemy a specific art of substance transmutation within a larger scientific and technical discipline, while Agricola talked about a generic art of metal as a separate activity from alchemy: "But concerning the art of alchemy, if it be an art, I will speak further elsewhere. I will now return to the art of mining."[7]

When talking about chemistry in the Renaissance we have to consider a variegated tradition of non-mechanical arts where chemical processes developed: pharmacy, metallurgy, glass making, dyeing, spirits distillation, agriculture cultivations and the preparation of colours for painting and drawing; each of these professional activities involve chemical processes that are not necessarily involved in alchemical transmutation. The difference between the purpose of chemical arts and alchemy was also the cause of cultural and epistemological tension during the Middle Ages and Renaissance. The medieval technological debate during the fifteenth and sixteenth centuries went beyond the traditional *Quaestio de alchimia* to assume an epistemological value in defining mechanical arts and in specific

informationem. [. . .] In metalllis vero ipsis informandis diligentior prorsus erit ratio adhibenda, ne, chimistarum elogio, operam simul et carbones perdidisse videamur. In qua quidem re summum erit metallorrum naturam cognoscere atque ubi deliquerint ebullierintque informare. Neque vero est quaod physiologica nunc ad ducam vosque ego doceam quae sit auri, argenti, aeris, stagni ac caeterorum natura, quidve ex horum confiat permixtura: equidem nolo noctuam Athenas aut videri chimicae sciens." (Gaurico, *De sculptura*, Book VI, ch. I, 1–14, ch. II, 1–8).

[2] On the debate about the meanings of 'chemistry' and 'alchemy' in the history of science, see Newman & Principe, "Alchemy vs. Chemistry"; and Abbri, "Chymists and Chymistry."

[3] Agricola, *De re metallica* (1950), XXV–XXXI.

[4] Cardano, *De rerum veritate*, Book X, ch. LI, 523–32.

[5] See Arnaud, *Introduction a la chymie*, 4.

[6] "tutti quelli c'hanno ragionato, o ragionano contra l'alchimia, e che tengono l'Alchimia in Arabico, o Chimia in Greco [. . .]" (Garzoni, *Piazza universale*, Book I, 139).

[7] Agricola, *De re metallica* (1950), XXIX.

artisanal processes of matter transformation. During the Middle Ages the alchemist protected the value of true alchemy, distancing it from artisan-alchemists who repeatedly failed in the transmutation process of metals, accusing them of being ignorant of the secret and true path of alchemy. At the end of the fifteenth and the beginning of the sixteenth century within the tradition of Italian technicians (engineers, architect, painters, etc.), practical and theoretical approaches to the studies of substances and their transformation produced and fired a debate about the truth and legitimacy of alchemy.[8] In the *Questione sull'alchimia* written by Benedetto Varchi (1503–1565) in 1544 in the entourage of Cosimo I de Medici (1519–1574) in Florence, the author discusses Biringuccio's position on alchemy within a debate usually reserved for the Scholastic tradition, giving credit for the first time to the opinion of a technician along with those of scholars and alchemists. Varchi considers Biringuccio's argumentation "irresolute and confused" but stresses how, in the end, he also gives a positive opinion on philosophical alchemy in contrast to sophistical alchemy:

> He spoke of alchemy in a very irresolute and inconclusive manner, as we see he does in his writings, where Several times he praises and blames it, but finally in the beginning of the ninth book He, no better and no worse than anyone else, confesses his hopes it could actually be possible after all, and cautions the men to exercise patience and have reverence for the magnificent goals, and miraculous works of nature. Although in the Chapter of Gold he so degraded Alchemy with very weak reasons, as one who had had a great deal of experience, and not a lot of science, since he did not even know that Alberto and many others had written of it. But still, we are obliged him for having been, beyond his great practical experience, a very loyal and truthful man, and most liberal of his treasures.[9]

The most important consequence of this inclusion of Biringuccio as a protagonist in the alchemical debate happens in the mid-sixteenth century; the *Quaestio de alchimia* can no longer exclude the position of the technician directly involved in the transformation of substances.[10] Leonardo da Vinci (1452–1519) and Biringuccio, the key figures in this article, were both involved in this debate and maturely and responsibly distanced themselves from the theoretical speculation of

[8] For Biringuccio and alchemy see, Perifano, "Alchimie," 189–202, *Alchimie à la Cour*; and Bernardoni, "Quaestio de alchimia." On the medieval alchemical debate, see Newman, "Technology and Alchemical Debate." On artisanal and vernacular epistemology, especially for the German area, see Smith, *Body of the Artisan*, 59–93, 142–51.

[9] "et in somma egli parlava dell'archimia molto confuso, et inresolutissimo come si vede ancora che egli fa ne' suoi scritti, dove molte volte la loda, e molte la biasima, ma finalmente nel principio del nono libro anch'egli, come tutti gli altri, per non essere forse né più di loro, né da meno, non che gli confessi in verità, che la sia possibile del tutto, ma conforta gli uomini ad esercitarla in reverenza per le stupende prove, e miracolose pera di lei; nonostante che nel Capitolo dell'Oro l'avesse tanto avvilita, con ragioni assai deboli, come quegli, il quale avea molta pratica, e non molta scienza, poiché egli non sapeva, lasciamo stare degli altri, ma che né Alberto ancora n'avesse scritto; ma comunque sia, gli semo obbligati grandissimamente essendo stato, oltre la grandissima pratica, uomo molto leale e veritiero, e liberalissimo dei suoi tesori." (Varchi, *Questione sull'alchimia*, 63–4).

[10] Bernardoni, *Conoscenza del fare*, 31–3.

alchemy and from uncontrolled techniques. Both of these authors gave credit to alchemy for producing codified procedures to transform substances and, although it is extremely difficult to talk about their role in the development of chemical technology, we do not have any doubts about their important contribution as protagonists and witnesses of the epistemological and technological transformations in the field of 'science and technology of matter'.

If we focus our attention on all the artisanal processes of substance transformation known and practiced during the fifteenth and sixteenth centuries it is easy to show that they do not fit perfectly within the domain traced by alchemy. This epistemological tension is usually neutralized by saying that alchemical and metallurgical disciplines are intricately interwoven and alchemists were involved in the development of metallurgy.[11] If one supposes chemistry did not exist as an independent and autonomous discipline before the end of the seventeenth century then speaking on this topic from a historical perspective in an Italian technical cultural context, some 'superior artisans' or 'artist-engineers' can be seen as having distanced themselves from alchemy by relocating its technology inside a new epistemological context in which chemical arts were developed and codified separately.[12] This context applies not only to metallurgical assaying but, more generally, it defines the area of the 'perfective arts', including all arts that use fire to work and transform matter.[13] In this *trading zone*, where Latin was scarcely known and where competition with alchemists was felt more keenly, a sub-group of artisans, artists and engineers kept their distance from alchemy to define a new field of 'chemical technology'.[14]

Artisanal processes could be seen as a sort of 'melting pot' of interwoven practical and theoretical knowledge coming from the world of crafts, including alchemy. It is prevalently in the world of artisans that we find references to 'chemical activity' during the sixteenth and seventeenth centuries. Before there were signs of a laboratory as a separate place—the most ancient plans to build a chemical laboratory are in the *Astronomiae instauratae mechanica* (Wandsbek, 1598) by Tycho Brahe (1546–1601) and in the *Commentariorum alchymiae* (Frankfurt, 1606) by Andreas Libavius (1560–1616), 'chemical operations' could be found everywhere 'perfective arts' were practiced.[15] This was in alchemist's laboratories, such as the one in Schloss Oberstockstall (Austria) in the sixteenth century, and assaying workshops, described in the anonymous *Probierbuchlein*

[11] Halleux, "Alchimiste et l'essayeur."; and Nummedal, *Alchemy and Authority*, 85–91.

[12] For more on the technical cultural context in the history of science and technology, see Zilsel, *Social Origins of Modern Science*; Maccagni, "Leggere, scrivere e disegnare"; Galluzzi, "Portraits of Machines; Smith, *Body of the Artisan*; Halleux, *Savoir de la main*, 102–39; and Long, *Artisan/Practitioners*, 10–29.

[13] Newman, *Promethean Ambitions,* 17–20.

[14] Long, *Artisan/Practitioners*, 94–6. See also, Galison, "Trading Zone."

[15] Hannaway, "Laboratory Design."

(1513), in Biringuccio's *Pirotechnia*, and in Agricola's *De re metallica*.[16] Further nascent signs of a distinct and autonomous chemical laboratory with purpose-built 'chemical' equipment can be found in the more general context of the Renaissance artist's workshop culture. Here, together with the fine arts such as painting, sculpture, architecture and jewelry making, metal carpentry, and gun casting, we find operations that use substance transformation, such as distillation and sublimation.

There is evidence that even the most important names in the history of art were involved in such workshop activities, for instance Michelangelo (1475–1564), whose Florentine workshop, as mentioned in his correspondence with his brother, was involved in generic metal carpentry such as the casting and welding of damaged swords and other craft objects.[17] Further evidence of artists' versatile activities during the Renaissance can be found in Andrea Verrocchio's (1435–1488) workshop, which specialized in painting, sculpture, casting and metal carpentry. One of the most important works created in this workshop when Leonardo was still one of Verrocchio's assistants was the copper sphere on the top of the lantern on Filippo Brunelleschi's (1377–1446) dome.[18] In Verrocchio's workshop, assistants and pupils, including Leonardo, prepared colors, glues, solvents, waxes, acids, alloys and so on. Being an artist in the Renaissance meant being part of a wider shared material culture as artisans and chymists used and developed the same materials and techniques.

In Biringuccio's book we find a description of a brass-making workshop he visited in Milan that discusses the organization of the workshop and the artisans' awareness of the technological and epistemological value of their works:

> [...] So that whoever entered that shop and saw the activity of so many persons would, I think, believe as I did that he had entered an Inferno, nay, on the contrary, a Paradise, where there was a mirror in which sparkled all the beauty of genius and the power of art.[19]

Biringuccio underlined the frenetic and very organized work in the workshop stressing the beauty of the mind of the artisans and the power of art. The harmonious cooperation between artisans and the plain consciousness of the processes of the art transformed the hell of hard and dirty metallurgical work into the heaven of the pursuit of the artistic goal. This oxymoron referred to the infernal and paradisiacal condition of art, exalting the extremely difficult process in which raw material is transformed into artifact, giving back an important social value to manual work.

---

[16] For a recent description of the chemical utensils found in the Schloss Oberstockstall laboratory, see Martinón-Torres, "Tools of Chymist"; "Probierbüchlein," 166–9; Biringuccio, *De la pirotechnia* (1977) 47r; and Agricola, *De re metallica* (1556), 174–208.

[17] Michelangelo, *Carteggio*, vol. I, 20–43.

[18] The sphere was three metres in diameter and was composed of several copper gores, soldered together on the top of the dome with a solar welder. For studies on the dome of Florence and the possible involvement of Leonardo see, Scaglia, "Studi tecnologici di Leonardo," 6–16; Di Pasquale, "Machinery of Construction Site"; and Galluzzi, *Mechanical Marvels*, 18–25, 99–116.

[19] Biringuccio, *Pirotechnia* (1942), 72. For the Italian, see Biringuccio, *De la pirotechnia* (1977), fol. 20r.

In the Renaissance technical tradition the *bottega* (workshop) was a place for work and not yet a space to conduct modern scientific research. Research on technology and natural phenomena was carried out thanks to the curiosity and the free enterprise of each individual artist; workshop mentality was conservative and technical knowledge was usually applied to technologically consolidated goals. The workshop was a place to produce specific artistic, technological and pharmacological objects yet it offered artisans the tools and the experiences to venture beyond the 'normalized activity'. We know several cases of painters involved in alchemy, such as Parmigianino (1503–1540), Cosimo Rosselli (d.1578), Lorenzo Lotto (1480–1556/7) and Domenico Beccafumi (1484–1551).[20] From the latter we have a series of engravings that represent some aspects of 'chem-alchemical' laboratory activity: heating systems, smelting furnaces, cucurbits, alembics and balanced scales (Fig. 1).[21] Alchemists used the same technological apparatuses and devices as artisans and alchemical research was often carried out along with chemical and metallurgical activities, for instance, at the *Fonderia* by the Medici family in Florence, or several other places in sixteenth- and seventeenth-century Europe.[22]

Leonardo and the 'Chemical' Arts

Very rarely have historians seen Leonardo from the perspective of his direct involvement in the chemicals arts. Apart from Ladislao Reti and a few others, Leonardo's observations and studies on matter transformation are usually seen in the context of his natural philosophy, with few references to his practical involvement as a painter and engineer during his entire career, including his vision and relationship with occult science and alchemy.[23]

If we focus our attention on Leonardo's notes and drawings about substance transformation, he can be seen as a witness and a protagonist of the epistemological

---

[20] For more information on the relation between artists and alchemy during the Renaissance, see Principe & DeWitt, *Transmutations*; Newman, *Promethean Ambitions*, 115–63; Ferino-Pagden et al., *Parmigianino*, 15–7; Calvesi, *Arte e alchimia*; Lennep, *L'art alchimique*; and Conticelli, *Alchimia e le arti*, 13–33.

[21] Gabriele, *Incisioni alchemico-metallurgiche*.

[22] On the Medici's *Fonderia*, see this volume: Kieffer and Beretta. On the relation of alchemy and metallurgical industry, see Nummedal, *Alchemy and Authority*, 119–46.

[23] For Leonardo's scientific and philosophical studies of natural phenomena and his theory of matter, see Boni, *Vinci e l'alchimia*, 401–5; Hooykaas, "Théorie corpusculaire de Léonard"; Gombrich, "Leonardo and the Magicians"; Vasoli, *Leonardo e l'alchimia*, 69–77; Kiang, "Leonardo and Alchemy"; Frosini, "Pittura come filosofia"; Kemp, *Leonardo da Vinci*, 293–329; Galluzzi, "Vinci's Concept of 'Nature',"; Beretta, "Leonardo and Lucretius"; Bernardoni, "Elementi, sostanze naturali"; and Nanni, "Lucrezio." For Leonardo's involvement in the chemical arts, see Schneider, *Chemische Wissen Leonardo*, 40–3, 87–92; Partington, *History of Chemistry*, vol. II, 1–8; Taylor, "Vinci et la chimie"; Reti, "Arti chimiche di Leonardo" (1952a; 1952b); Bernardoni, "Leonardo and the 'chemical arts'," and *Esplosioni, fusioni e trasmutazioni*, 38–41.

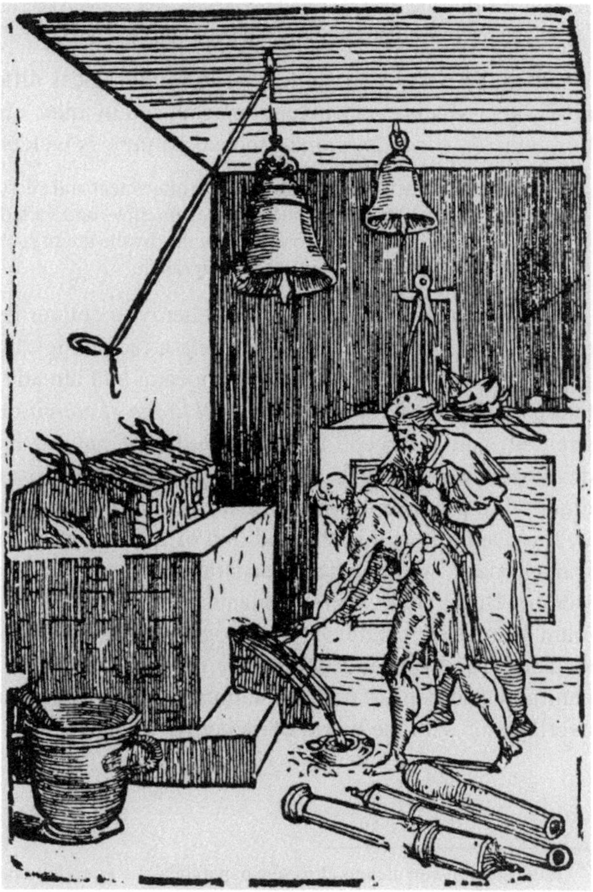

Fig. 1 D. Beccafumi, Foundry workshop; courtesy of Accademia Carrara, Bergamo

tension between mechanical arts and alchemy during the Renaissance period. His attention was specifically focused on alchemy's claims to overturn the relationship between art and nature; however, he also stressed that alchemy was worthy of pursuit as it might produce knowledge that could improve the condition of mankind:

> Man is involved with things produced by nature and nature does not change the ordinary kinds of things it creates in the same way that from time to time the things created by man are changed; and indeed man is nature's chief instrument, because nature is concerned only with the production of elementary things, but from these elementary things man produces an infinite number of compounds, although he has no power to create any natural things except another like himself, that is his children. [...] And of this the old alchemists will serve as my witnesses, who have never either by chance or deliberate experiment succeeded in creating the smallest thing which can be created by nature; [...] If, however, insensate avarice should drive you into such error, why do you not go to the mines where nature produces this gold, and there become her disciple? She will completely cure you of your

> folly by showing you that nothing which you employ in your furnace will be numbered among the things she employs in order to produce this gold.[24]

After the acknowledgement of the ontological difference between nature and artificial products every invention produced by man, elsewhere significantly called by Leonardo *seconda natura* (second nature), is welcome:

> Gravity and force together with material movement and percussion are the four accidental powers by which the human race in its marvellous and varied works seems to reveal itself and a second nature in this world; seeing that by the use of such powers all the visible works of mortals have their existence and their death.[25]

He argued that the inability of alchemy to deliver its claims of transmutation is due to the impossibility of developing a technology able to reproduce the natural genesis of substantial forms. As Avicenna had already claimed in his Latin translation of *Kitab ash-Shifa'* (*De congelatione et conglutinatione lapidum*), Leonardo stressed man's perception of the purity of natural elements was neither highly developed nor sensitive enough to be able to precisely quantify the proportions required to produce the substantial form of different metals.[26]

If we assume within the tradition of Italian 'artist-engineers' this 'epistemological approach' was oriented to separate true and false alchemy and culminated in the work of Biringuccio, then the manuscripts of Leonardo could be seen as one of the main sources of 'chemical technologies' from the end of fifteenth and beginning of sixteenth centuries and one of the most important sources for the studies of the natural phenomena of transformation in a cultural context separate but in many case overlapping with the world of alchemy.

---

[24] "E questo non è in alcuno altro senso, perché sol s'astendono nelle cose che al continuo produce la natura, la qual non varia le ordinarie spezie delle cose da lei create, come si variano di tempo in tempo le cose create dall'omo, massimo strumento di natura. Perché la natura sol s'astende alla produzion de' semplici, ma l'omo con tali semplici, produce infiniti composti ma non ha potestà di creare nessun semplice se non un altro se medesimo, cioè li sua figlioli. E di questo mi saran testimoni li vecchi archimisti, li quali mai, o caso o con volontaria sperienzia, s'abbatterono a creare la minima cosa che crear si possa da essa natura [...]. E, se pur la stolta avarizia in tale errore t'invia, perché non vai alle miniere dove la natura genera tale oro e quivi ti fa suo discepolo, la qual fedelmente ti guarirà della tua stoltizia mostrandoti come nessuna cosa da te operata nel foco non sarà nessuna di quelle che natura adopri al generare esso oro." (Leonardo, *Corpus degli studi anatomici,* fol. 50v [19045v]. Translated from Leonardo, *Corpus of Anatomical Studies,* vol. I, fol. 50v [19045v]).

[25] "La gravità, la forza insieme col moto materiale e lla percussione sono le quattro potentie accidentali colle quali l'umana spetie nelle sue mirabili e varie operationi pare in questo mondo dimostrarsi una seconda natura. Imperoché con tali potentie tutte le evidenti opera de' mortali ànno loro essere e loro morte." (Leonardo, *Codex Arundel,* fol. 151v). See Frosini, "Forza in Leonardo da Vinci," 121.

[26] These admonitions against alchemy, presented for the first time by Avicenna, were known during the Middle Ages as *sciant* artifices, see Avicenna, *De congelatione et conglutinatione lapidum,* 53–5; Newman, "Technology and Alchemical Debate."; and Halleux, "Alchimiste et l'essayeur," 211.

Leonardo's manuscripts are one of the best sources to study the chemical equipment of Renaissance. His tireless sketching during his entire career produced a very large and uncommon collection of manuscripts that allow us to understand and visualize many tools, instruments and chemical processes studied, developed and used during the Renaissance.

An activity that was a sort of flywheel in involving Leonardo in the study of the many aspects of bronze casting was the project of the *Equestrian monument for Francesco Sforza*. This was a very large bronze statue that was never cast but studied and worked on by Leonardo for almost 20 years. The main difficulty of this cast was the size, more than 7 m tall and almost 70 tons in weight, and the plan to realize it in a single pouring. To organize the foundry and the molding process, Leonardo studied artillery casting, furnaces and the materials used in the molding, like sand, wax, gypsum, clay, brick powder, carrying out several experiments to better understand their chemical-physical properties.[27]

Leonardo's manuscripts are among the most important sources for the artillery production process, surely the most detailed and best illustrated of the fifteenth and sixteenth centuries. In some sheets of the *Codex Atlanticus*, for instance, we may find the first and only reference for the production of iron shaft-soldering bombards, such as the famous Belgian Mons Meg, created with a hammer at the forge. To minimize the inaccurate results of forge hammering he planned a draw bench machine to make homogeneous iron shafts in order to simplify the welding and produce stronger cannons (fol. 10r, 11r, 15v, 41r).

Several studies are dedicated to the casting process of bronze bombards.[28] Leonardo's manuscripts are once again the most ancient source which make it possible to visualize the several phases of the molding process both for the *tromba* (chase) and the *coda* (breech), the two parts of a gun which were assembled by threaded coupling. Leonardo gives us very detailed drawings, such as the channel for pouring the bronze into the mold during casting (*Codex Atlanticus*, fol. 46r, 53r, 60r, 61r, 937v) (Fig. 2).

Leonardo's studies on reverberatory furnaces is one of the best examples to demonstrate that his mental approach to technological problems went beyond the creation of a specific process or device. His detailed drawings of furnaces based on very close observation allowed him to let his curiosity take him beyond the improvement of the technical apparatus towards some very interesting observations and considerations on the nature of fire and its penetrative power.

Folio 87r of the *Codex Atlanticus* portrays several kinds of reverberatory furnaces and among them there is a drawing of a curious and obscure apparatus. The drawing portrays a fusion chamber above a very tall platform connected to a firebox by two vertical ducts. On the left there is also another duct that probably conducts

[27] Bernardoni, "Leonardo and the Equestrian Monument"; and Brugnoli, "Scultura di Leonardo."

[28] Bernardoni, *Esplosioni, fusioni e trasmutazioni*, 26–35; Brioist, *Vinci, Homme de guerre*, 105–14.

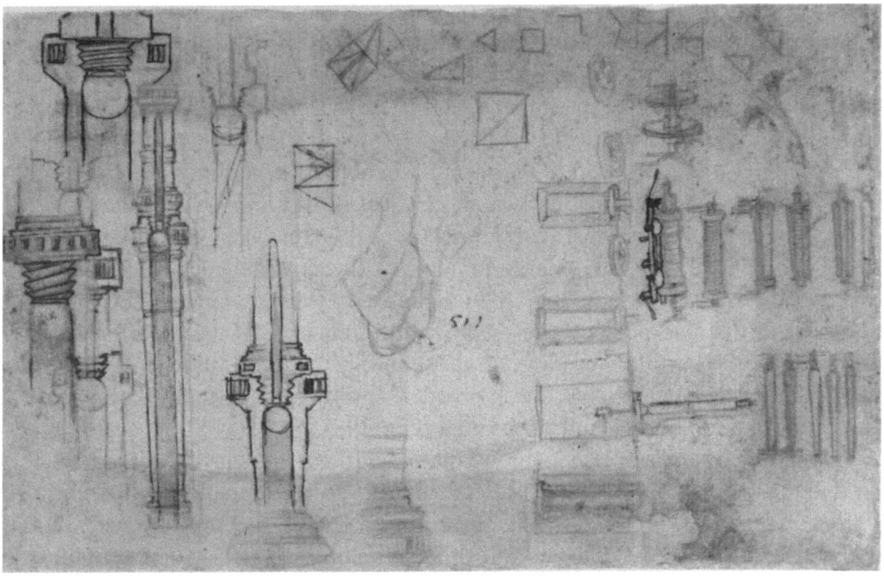

Fig. 2 Leonardo, artillery moulding process for the barrel, Codex Atlanticus, fol. 46br, Biblioteca Ambrosiana, Milan (Image taken from Codex Atlanticus, Hoepli edition, Milan, 1894–1904)

air into the chamber from below. The note under the drawing helps us interpret the device as a study for amplifying the penetrative action of fire: "The greater the natural motion of the fire or the greater its weight, the greater its impulsive force" (Fig. 3).[29] Leonardo talks about fire in the same terms as the weight of bodies, so he interprets it as a hard material particulate flow, able to penetrate the body and break the link of its particles. In Aristotelian matter theory, fire is the lightest element that finds its natural place at the most peripheral region of the sub-lunar world. Just as a stone falls towards the earth to reach its natural place, fire goes in the opposite direction, rising to the far reaches of the sky; the longer the distance, the quicker it rises. So towards the vertical ducts of these experimental furnaces, the particles of fire would have to increase their impact speed on the mass of metal inside the fusion chamber. The short sentence below the drawing of the furnace, along with many other observations of the physical status of the four elements and other natural substances, enables us to reconstruct Leonardo's theory of matter as a sort of a 'Aristotelian corpuscolarism' and a kinetic theory of heat/fire in which the melting of metal or the dissolution of a substance by fire depends on the percussion of a flow of particles.[30] This interpretation finds confirmation in several reverberatory furnace drawings made with bent surfaces to drive the flow of fire into the middle of the fusion chamber (*Codex Atlanticus*, fol. 1103r, 548r, 580v, 82v) (Fig. 4).

[29] "quanto più il moto natural del foco o del peso sia lungo, più vale la sua percussion." (Leonardo, *Codex Atlanticus*, fol. 87r).

[30] Bernardoni, "Elementi, sostanze naturali."

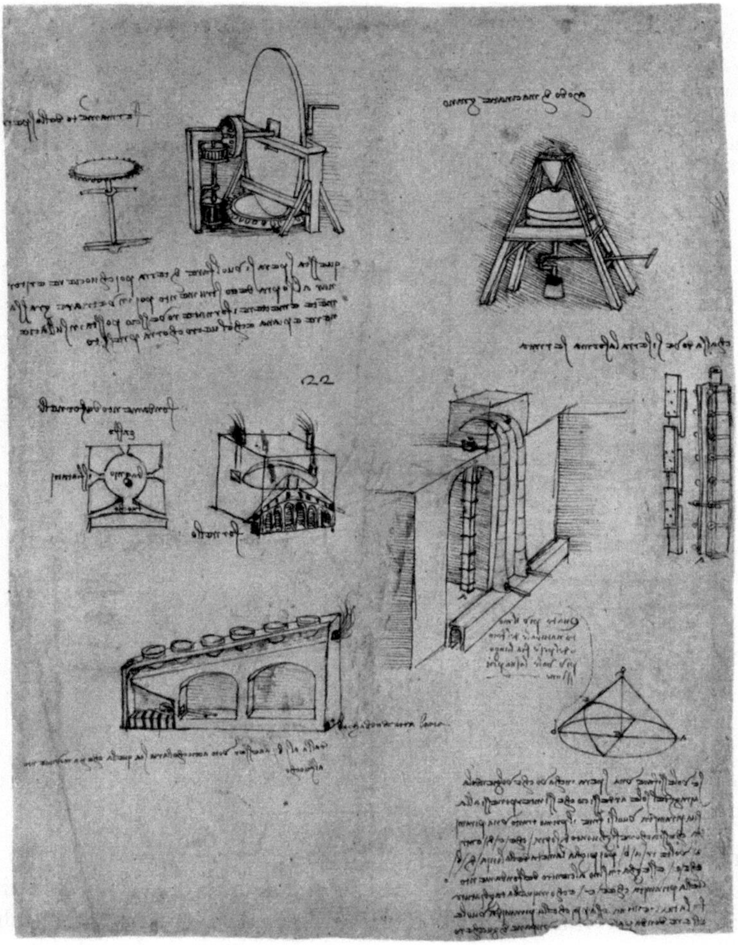

Fig. 3 Leonardo, Technological application of the element of fire, Codex Atlanticus, fol. 87r, Biblioteca Ambrosiana, Milan (Image taken from Codex Atlanticus, Hoepli edition, Milan, 1894–1904)

Furnaces become Leonardo's instrument for studying and observing fire. A drawing on a folio in the *Codex Arundel* representing a tower furnace, used in the Middle Ages by alchemists for distillation, assumes a very important epistemological value (Fig. 5). The drawing presents an apparently accurate copy of this furnace's vertical section and, when we read the note below, we understand that his interest is not in the technological device but the transformation of the elements involved in the combustion process inside it. The note below the furnace helps us understand that

> once you have dealt with the motion of heavy solids, deal with heavy liquids and with air and with the motion of fire. Compare the motion of fire with the whirls of air and water and

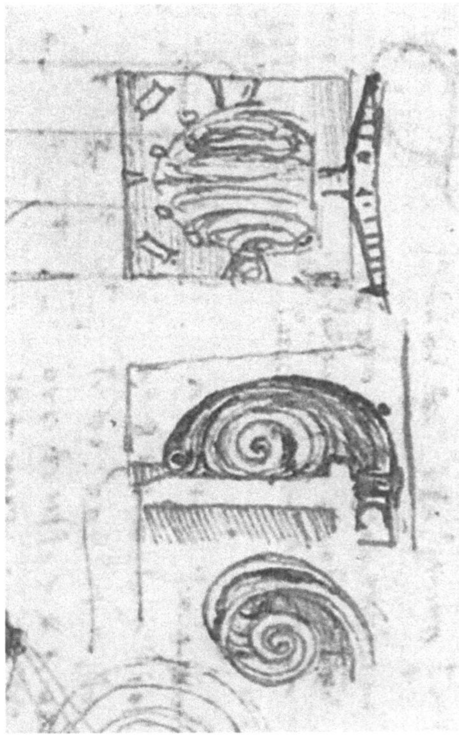

Fig. 4 Leonardo, Vortices of fire, Codex Atlanticus, fol. 580v, Biblioteca Ambrosiana, Milan (Image taken from Codex Atlanticus, Hoepli edition, Milan, 1894–1904)

you will find a drilling motion of fire that makes it powerful for fusion; you can obtain these gyrations with the help of registers and boiling water.[31]

Leonardo's main interest in this drawing is to create a device for him to see the drilling motion of fire. Leonardo's goals are both technological and scientific; the fire works as a drill. If this were to be verified, it could be used in some technological application but it could also assume a more general scientific and philosophical explanation because the power of fire, like the power of the other elements (water, air and earth), manifests itself in the form of a spiral—the natural screw. The power of the screw could be seen in the vortices of water and air, but also in Brunelleschi's bigger cranes and powerful technological elevator devices that become an analogic explanation for the forces operating in nature.[32]

Another very interesting case involving the chemical equipment studied by Leonardo is the development of the alembic refrigerator system. In the *Codex Atlanticus* there are some drawings (fol. 912r, 1114r a–b, 216r, 989r, 1118r)

[31] "Trattato che ài de' moti de' solidi gravi, trata de' gravi liquidi e dell'aria e de' moti del foco, e col moto di questo foco fa conparatione del moto delle revertigine dell'aria e dell'acqua, e troverrai moti trivellanti del foco a ffarlo potente alle fusioni colle sua revolutioni, la qual cosa farai co' regisstri e con acqua bollente." (Leonardo, *Codex Arundel*, fol. 145v).

[32] Pedretti, *Leonardo architetto*, 9–12; and Galluzzi, *Mechanical Marvels*, 55–6.

Fig. 5 Leonardo, Drilling motion of fire, Codex Arundel, fol. 145v, Courtesy of British Library, London

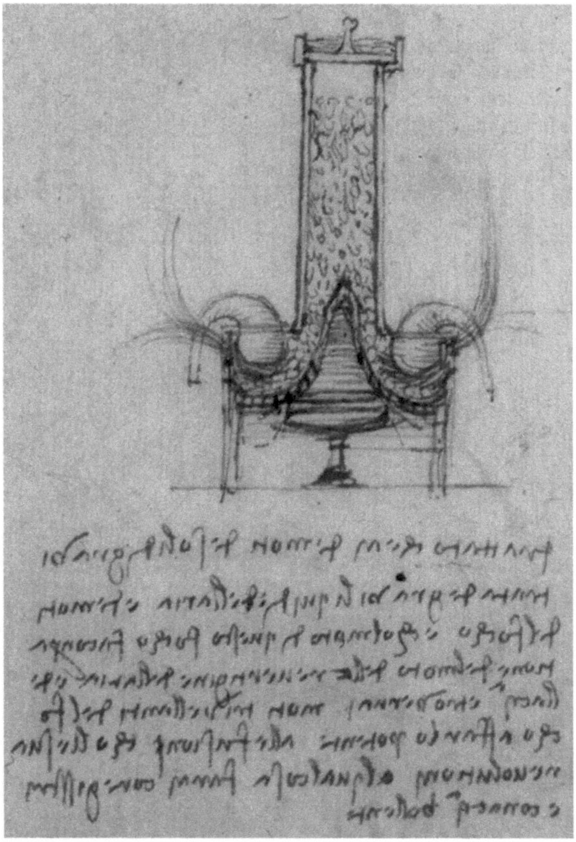

where we can see a very interesting attempt to develop a basin refrigerator system by separating the basin and the stove to prevent a thermal shock in the neck of the cucurbit (fol. 989r) (Fig. 6). For the refrigerator system Leonardo also developed an alembic lid with a chamber along its external surface where fresh water could flow inside (Fig. 7). Scholars had already discussed this very modern solution in order to understand its pertinence to a real alembic or to an unrealized design.[33] The wireframe representation leads us to think that it was made of glass, but its morphological complexity tends to suggest it was made by hammering soldered copper.

In folio 216r of *Codex Atlanticus* we have another interesting study for casting a traditional alembic (Fig. 8). This is the only source known to me with specifications about creating and conserving an alembic. First of all, writes Leonardo, the model of the alembic in turned wood has to be built, and then it has to be refined with clay wool-cloth clippings. The external plaster mold has to be built on it in two valves.

[33] Reti, "Arti chimiche di Leonardo" (1952a; 1952b).

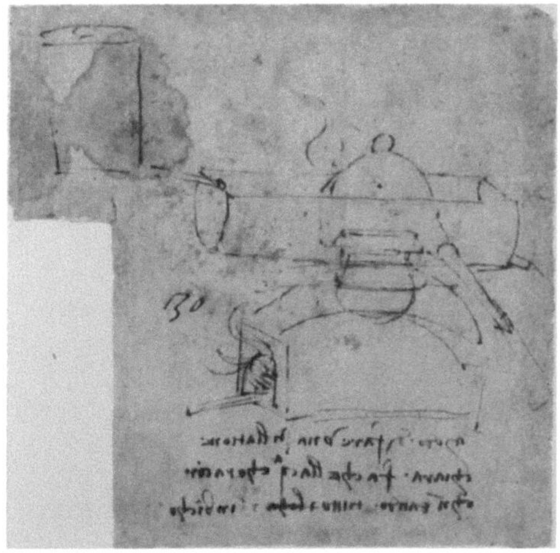

Fig. 6 Leonardo, Basin refrigerator alembic, Codex Atlanticus, fol. 989r, Biblioteca Ambrosiana, Milan (Image taken from Codex Atlanticus, Hoepli edition, Milan, 1894–1904)

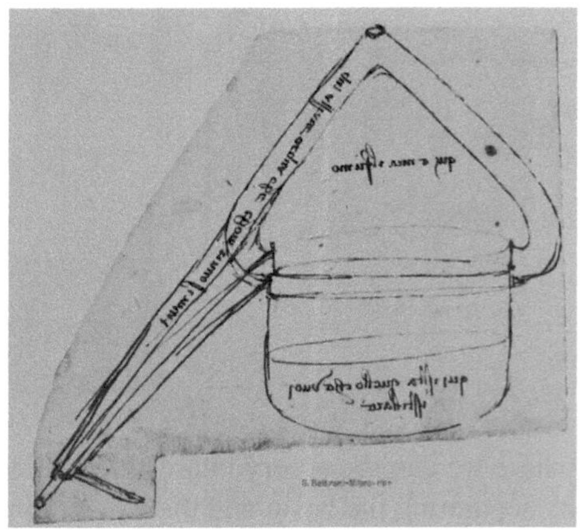

Fig. 7 Leonardo, Refrigerated Alembic, Codex Atlanticus, fol. 1114br, Biblioteca Ambrosiana, Milan (Image taken from Codex Atlanticus, Hoepli edition, Milan, 1894–1904)

There is no other information and it is impossible to know the casting material Leonardo thought he would use. It was most likely copper or bronze, however, we cannot exclude the use of glass. Leonardo, in fact, preferred using glass alembics because of the possibility of observing condensation phenomena inside them. In a fragmentary sheet of *Manuscript E* he specifies:

> Such is the nature of the condensation of the walls, constraining the space enclosed between them, as is that of the enclosed [space], multiplied by the enclosing [walls]. This is proved with the smoke generated in an enclosed space, as is seen in the glass vessels with which distilling is done, in which it is easily recognized at what part of this transparent vase the smoke condenses more or less and . . .[34]

[34] Leonardo, *Manuscript E*, 8 [fol. 3r, I].

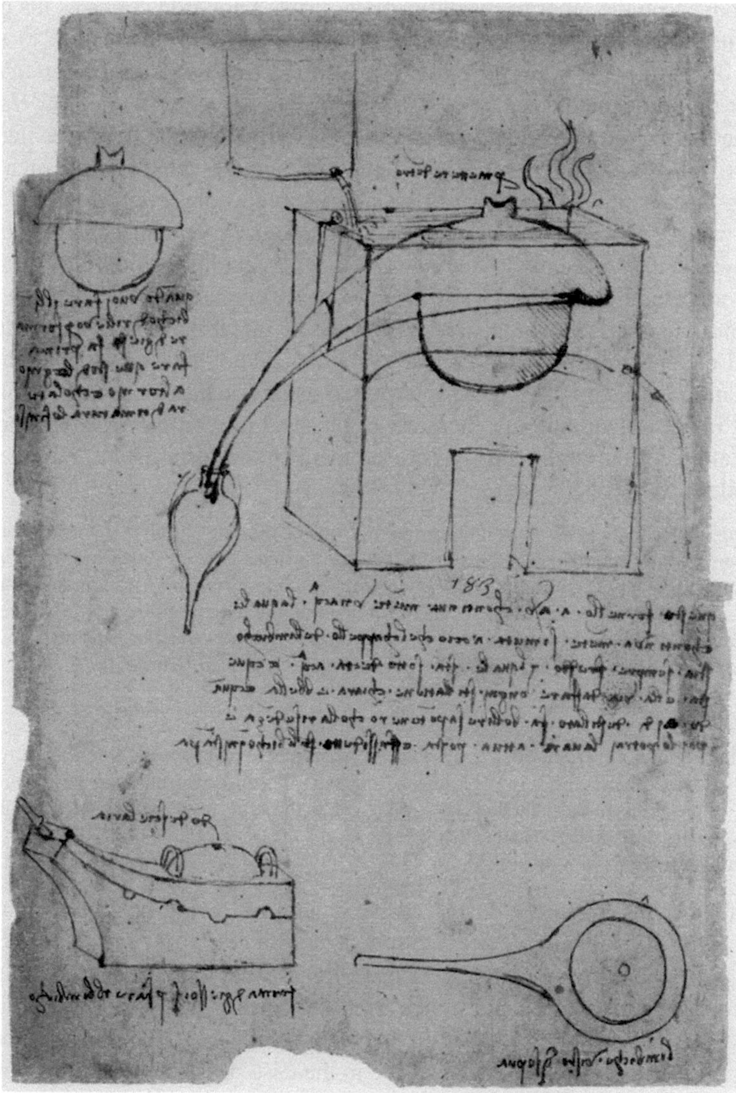

Fig. 8 Leonardo, Basin refrigerator alembic, Codex Atlanticus, fol. 216r, Biblioteca Ambrosiana, Milan (Image taken from Codex Atlanticus, Hoepli edition, Milan, 1894–1904)

The use of glass in the casting is very difficult because of its viscosity. The only reference to this technique in Leonardo is in relation to the casting of a glass bell jar in *Manuscript B*, in which he says he uses a metallurgical furnace to pour the glass inside a very hot mold: "Prepare the furnace in the usual manner, that is, for bombards, and when the glass is melted, pour it into the red–hot form."[35]

[35] Leonardo, *Manuscript B*, 19 [fol. 10v, IV].

Moreover, on the other pages of his manuscripts Leonardo confirms his propensity to use glass for various apparatuses such as the transparent box to see the motion of the flowing water through a hole in the bottom (*Codex Atlanticus*, fol. 219r) or motion generated in a mass of water by a surface wave (*Codex Hammer*, fol. 9r and 29v). We also find a transparent glass cylinder for studying the flame of a candle (*Codex Atlanticus*, fol. 226r) and, most of all, the drawing of a glass-drinking horn with which Leonardo says he is able to observe free atoms moved by the water.[36] The topic of atoms in Leonardo is a very articulated question, interwoven with the theme of the *essere del nulla* (being of nothing), for which there is a specific literature.[37] Here it is enough to make a brief reference to this theme, just to stress it could also be discussed in terms of the observation of physical-chemical phenomena. Leonardo's atoms do not have qualitative-chemical specifications but they have to be interpreted as physical primary matter particles existing in nature without any chemical or physical properties. They are the result of substance consumption and we can see them in the form of dust, smoke, or as tiny, undefined pieces of matter:

> The air that successively surrounds a moving object moving through it makes various motions within itself. This can be seen in the dust particles [*attimi*] found in the sphere of the sun, when they penetrate through some window into an obscure location, and when a stone is thrown into these dust particles [*attimi*], along the length of this solar ray, you can see the dust particles [*attimi*] turning about at that location where the path made by the moving object was filled in again by the air, as has been proved in the fifth.[38]

And again:

> It follows therefore, from what I say, that the atmosphere acquires its blueness from the particles which catch the luminous rays of the sun. We may also observe the difference between the atoms of dust and those of smoke seen in the sun's rays as they pass through the chinks of the walls in dark rooms, that the one seems the color of ashes, and the other—the thin smoke—seems of a most beautiful blue.[39]

Each substance could be destroyed by separation, going beyond the limits of *minima naturalia*—the last specification of substantial form on matter. Significantly, even though the sentence is partially crossed out, Leonardo writes the "atoms are not a part of the substance from which they are born," as he wants to take under consideration the possibility of the strangeness of the inner composition of substance.[40] Atoms as we will also see in the case of Biringuccio are something

[36] "Vetro a ciò che si vegga li attimi nell'acqua che si move." (Leonardo, *Codex Atlanticus*, fol. 589v).

[37] Marinoni, "L'Essere del nulla," 209–32; Bernardoni, "Elementi, sostanze naturali," 99–104; Beretta, "Leonardo and Lucretius"; and Nanni, "Lucrezio."

[38] Leonardo, *Manuscript F*, 138 [fol. 74v, IV]. The phenomenon is also noted on folio 87r, see Leonardo, *Manuscript F*, 158–9 [fol. 87r]; and on folio 78r, see Leonardo, *Manuscript L*, 89–90 [fol. 78r].

[39] Leonardo, *Codex Hammer*, 4A, fol. 4r.

[40] "attimo non è parte della materia donde nascie." (Leonardo, *Codex Arundel*, fol. 176v).

material, which could be seen or imagined observing natural or artificial phenomena of matter transformation.

Biringuccio's Pirotechnia and the Knowledge of Nature

Biringuccio, like Leonardo, placed the focus of his consideration on the possibility of increasing knowledge and reproducing processes of nature through the technological codification emerging from alchemical tradition and from his professional activity. In the *Pirotechnia* (Fig. 9), we again find the two traditional contradictory opinions on alchemy: negative towards the alchemists who hide their art behind esotericism, magic, alchemical authority and those who practice chemical process without ratio and empirical control, and positive in relation to the improvement of mankind:

> How many alchemists have I heard lamenting, one because by some unfortunate chance he had spilled his whole composition in the ashes; another because he had been deceived by the excessive strength of the fire, so that the substance of his materials had been burned and the spirits inadvertently allowed to escape; and yet another because he had poor and feeble materials! In a word, one for one reason, and one for another, in order to hide either their deception or their ignorance, all defend themselves and make excuses for their art.[41]

Moreover, after marking his distance from false and sophistic alchemy, Biringuccio returned to what he called "true alchemy" with very positive opinions, talking about it as a philosophical and technological activity grounded on reason and empirical experimentation which, even when following the illusory goal of metal transmutation or the perfective and healthy elixir, discovered new substances, medicine and technological processes. After a rational and empirical codification of the chemical process and substance discovered and prepared by it, true alchemy remained to Biringuccio the most important "art" for studying the secrets of nature:

> Thus, in short, it can be said in conclusion that this art [alchemy] is the origin and foundation of many other arts, wherefore it should be held in reverence and practiced. But he who practices it must be ignorant neither of cause nor of natural effects, and not too poor to support the expense. Neither should he do it from avarice, but only in order to enjoy the fine fruits of its effects and the knowledge of them, and that pleasing novelty which it shows to the experimenter in operation.[42]

Biringuccio considered alchemy a legitimate knowledge grounded on experimental and speculative activity. This is an important epistemological step for artisanal processes because alchemy was no longer seen as a utilitarian art but as a disinterested activity carried out by men to enjoy the fruits of their knowledge.

---

[41] Biringuccio, *Pirotechnia* (1942), 41. For the Italian, see Biringuccio, *De la pirotechnia* (1977), fol. 7v.

[42] Biringuccio, *Pirotechnia* (1942), 337. For the Italian, see Biringuccio, *De la pirotechnia* (1977), fol. 123v.

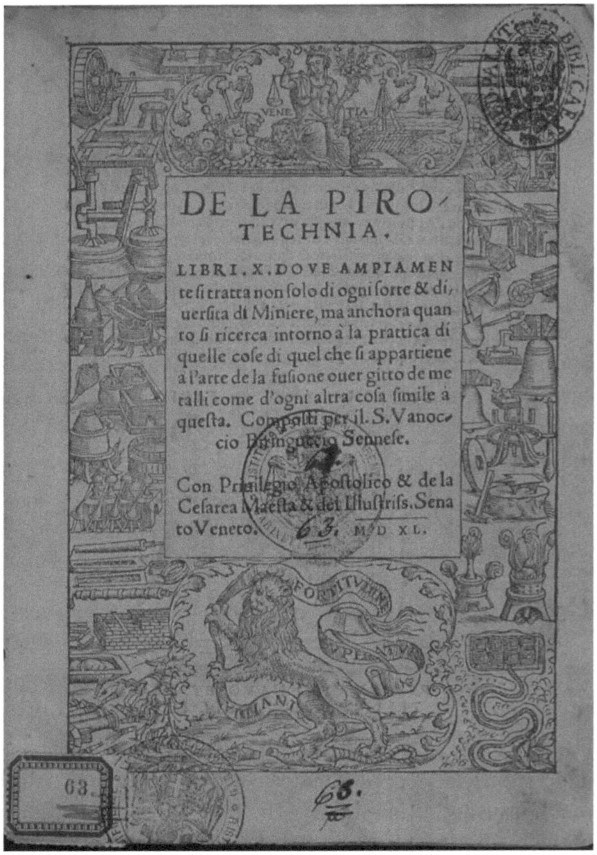
DE LA PIRO-
TECHNIA.
LIBRI.X.DOVE AMPIAMEN
te ſi tratta non ſolo di ogni ſorte & di-
uerſita di Miniere, ma anchora quan
to ſi ricerca intorno à la prattica di
quelle coſe di quel che ſi appartiene
à l'arte de la fuſione ouer gitto de me
talli come d'ogni altra coſa ſimile à
queſta. Compoſti per il S. Vanoc-
cio Biringuccio Sennese.
Con Priuilegio Apoſtolico & de la
Ceſarea Maeſta & del Illuſtriſs. Sena
to Veneto. M.D.XL.

Fig. 9 V. Biringuccio, *De la pirotechnia*, Venice, 1540, frontispiece; courtesy of Museo Galileo Library

"Philosophical alchemy," as Biringuccio calls it, is for him a scientific discipline that directly studies natural phenomena and could be seen as an alternative approach to the theoretical speculation of scholastic philosophy.[43] Alchemy was here presented and discussed from a technical and theoretical point of view and, even more importantly, it was classified as a specific art of fire placed close to, but separate from, the arts of distillation. This is an important distinction from which Biringuccio's original approach to the arts of fire emerges and that distinguishes him, for instance, from Georg Agricola who, at almost the same time, wrote his *De re metallica* in which he outlined the general arts of metals related to mining and metallurgy. Biringuccio expanded his perspective to include all the arts of fire, including, artillery, making gunpowder, acids preparation, pottery, goldsmithing, fireworks, generic distillation arts, as well as alchemy.[44] If we accept his conditions,

[43] See Biringuccio, *De la pirotechnia* (1914), 27–40; Perifano, "Alchimie"; Smith, *Body of the Artisan*, 143–4; and Bernardoni, *La conoscenza del fare*, 669.

[44] Biringuccio, *De la pirotechnia* (1977), fol. 123r–4r.

alchemy and pyrotechnics are not in contradiction, rather, in so far as they are used in the same philosophical concepts and the same operational practices, alchemy could be considered a continuity of the arts of fire. Biringuccio's *Pirotechnia*, in fact, could be interpreted as a sort of normalized alchemy, that is, an 'engineering discipline' founded on a general theory of matter and in codified operative practices to control the power of fire during matter transformation processes.

The importance of Biringuccio's book in the history of metallurgy and chemistry is well known. Along with Agricola's works, it was a most important source for many chemical technology devices and processes.[45] Although *Pirotechnia* was essentially a practical treatise, the author's intention was to go beyond the writing down of *know how* towards a theoretical consideration about the nature and transformation of substances. As can be seen in many passages of the book, Biringuccio developed independent ideas on these issues, combining different positions on the mineral world from philosophical and alchemical traditions, in particular from Albertus Magnus's (1193–1280) *De mineralibus*, and the *Summa perfectionis* of Pseudo-Geber, but he was also influenced by the new concept of substantial form introduced by Augustine Nifo (1473–c.1538) in the Aristotelian tradition of Padua, who considered the essence of natural substances to be the order and internal structure of sub-particles.[46]

Biringuccio classified mineral substances on the basis of their macroscopic properties observable during metallurgical processes and starting from these he tried to give a description of metal into the frame of a general theory to explain their mutual physical-chemical relationship. His model of perfection was gold; its gloss, compact structure, resistance to oxidation and calcination made this the mineral substance with the most balanced internal particle structure:

> [...] Its original and peculiar materials are none other than elemental substances, with the quantity and quality of each proportioned equally one to the other and very finely purified. From this union of elements which are of equal force there is born a pleasing and perfect elemental mixture, and then after fermentation and decoction the elements finally become fixed, permanent, and joined together in such a union that they are almost inseparable, so that by the power of the heavens or of time or of the order of the most wise Nature, or by all these together, these substances are converted into that metallic body called gold.[47]

The other metals, starting with silver, have an internal growing disequilibrium that manifests itself by observing external physical-chemical properties. This classification becomes even more evident with the category of semi-minerals in which, for example, Biringuccio includes substances such as mercury, marcasite, sulphur and antimony that seem to be progressive stages of the metal generation process

[45] See Biringuccio, *De la pirotechnia* (1914), VII–XXV, and *De la pirotechnia* (1977), pp. I–XXXIII; and Cipriani, "Agricola e Biringuccio."

[46] Emerton, *Scientific Reinterpretation of Form*, 97–105; and Bernardoni, *La conoscenza del fare*, 71–114.

[47] Biringuccio, *Pirotechnia* (1942), 26–7. For the Italian, see Biringuccio, *De la pirotechnia* (1977), fol. 1r.

defined on the basis of their appearance and their chemical-physical behavior during empirical manipulation. The case of antimony is one of the best examples:

> [It] is a composition made by Nature to create a metallic mineral that is overflowing with an undue proportion of hot and dry material and with its moisture poorly mixed, with an effect wholly contrary to the composition of metals. Therefore, it comes to be, like quicksilver, a mineral deformity and monstrosity among metals. Or it might be a material that is about to reach metallic perfection, but is hindered from doing so by being mined too soon. I am persuaded to this opinion by seeing in it so many parts similar to those of metals [. . .].[48]

As we have seen in Leonardo, Biringuccio also used specifically designed tools and metallurgical processes to observe natural phenomena and to underline the importance of empirical research.

Describing the several aspects of distillation technology, Biringuccio talked about the possibility of going beyond the physical dimension of the four elements in order to reach an undetermined status of matter, identified by the alchemist as the "quintessence":

> By means of the art of distillation many say that you proceed from element to element, making them subtle and separating them so that at the end the materials are reduced to such a point that they no longer have a resemblance to any of the substances of the four elements. And then they say that they have reduced them to one, which they call the Quintessence.[49]

Biringuccio did not believe that this physical status of matter could be identified with quintessence but it is very interesting to read the possibility of having a physical dimension of matter beyond the four-elementary specifications achieved through a technological process.

Biringuccio did not give an organic and systematic explanation of this status of matter but, as in the case of Leonardo, from his book arises a corpuscular concept of matter in which he spoke about atoms, giving examples of their physical dimension. Particularly interesting was his analogical explanation of the compactness and the regular size of the particle of gold as it is seen in the separating process separating silver and gold: silver turns into solution with nitric acid while gold remains behind as a solid particle: "It [goes] here and there, wandering about in the water like atoms. Because these are tiny and subtle things, they reduce the power of water."[50] Atoms are light, small and are driven by water in casual motion. Even though it is impossible to talk about atomistic conception of nature, it is very interesting to note how Biringuccio does not limit himself to technical considerations about metallurgical process but, uses his tools (cucurbit, alembic, furnaces etc.) to observe matter transformation processes in order to have a deeper understanding of natural and artificial substances.

---

[48] Biringuccio, *Pirotechnia* (1942), 91. For the Italian, see Biringuccio, *De la pirotechnia* (1977), fol. 27v.

[49] Biringuccio, *Pirotechnia* (1942), 340–1. For the Italian, see Biringuccio, *De la pirotechnia* (1977), fol. 125r.

[50] Biringuccio, *Pirotechnia* (1942), 200. For the Italian, see Biringuccio, *De la pirotechnia* (1977), fol. 71r.

The atomic dimension evoked by Biringuccio seems to involve the physical status of matter and, even though he did not have a clear position on it, his reference to the genesis of substances reveals to us his interest and involvement on the problem of generation and diversification of natural species:

> I am sure that you understand that of all the things created by the most high God Himself or by Nature at His command, not one—even though it be an atom or the smallest worm—has been produced without some particular gift.[51]

These incursions into problematic topics of natural philosophy are clear evidence of Biringuccio's willingness to participate in the debate on natural and technological phenomena. Apart from the question of how this would fit into the context of natural philosophy and the genesis of substance, it is evident that the problem of the composition of substances was a debated topic and Biringuccio is one of the best examples of this interest in the first half of the sixteenth century.[52] This concept of atoms—as opposed to the *minima naturalia* developed in the Aristotelian medieval tradition—seems to be close to the Democritean concept, even though Biringuccio, and Leonardo, rejected the mechanical explication of substance composition by continuing to talk of Aristotelian elements as a final part of substance constituents. As can be seen in the cases of Leonardo and Biringuccio, Renaissance technicians seem to consider atoms as pieces of undetermined matter, as something existing beyond the limits of the physical dimension that could not be reached by mechanical (by cutting) or chemical (by distillation) means. A very interesting partial confirmation of this comes from the curious assumption during the late Middle Ages that atoms were the last division of the official weight set. We find this in the writings of Antonio Averlino detto il Filarete (c.1400–c.1469), Girolamo Cardano and Gerolamo Cattaneo and atoms were used as units of linear measurement in several North Italian towns. This remained in place until the introduction of the modern metric system.[53]

Biringuccio's scepticism about alchemists and the bookish tradition drove him to stress the importance of factual verification in the advancement of learning as the true foundation of the knowledge of nature. Biringuccio's idea of knowledge as based on factual verification is to be situated in the technical culture of which he is part. To stress the importance of factual knowledge he underlines the importance of

---

[51] Biringuccio, *Pirotechnia* (1942), 114. For the Italian, see Biringuccio, *De la pirotechnia* (1977), fol. 36v.

[52] For a more detailed analysis of Biringuccio's concept of 'atoms' and his theory of matter, see Bernardoni, *La conoscenza del fare*, 78–105. For a general analysis on the medieval and Renaissance concept of 'matter', see Murdoch, "Minima Naturalia"; and Grellard & Robert, *Atomism*.

[53] Filarete, *Trattato di architettura*, 470; Cardano, *Practica arithmetica*, ch. LXIII; and Cattaneo, *Dell'arte del misurare*. For a detailed analysis of the units of measure use in Italy from the Middle Ages to the modern era, see Frangioni, *Metrologia lombarda*.

evidence from nature whatever its origin (books, nature, oral tradition) and concludes:

> I have done this willingly in order that you may acquire more learning and because I am certain that new information always gives birth in men's mind to new discoveries and so to further information. Indeed I am certain that it is the key that arouses intelligent men and makes them, if they wish, arrive at certain conclusions that they could not have reached without such a foundation, or even nearly approached.[54]

The advancement of learning for Biringuccio depends on continual "discoveries" and "new information," that results in two processes of knowledge: the first leads to the creation of new artificial products and the second determines the increase in and deepening of knowledge about nature. As is clear from this passage, human creativity can be awakened by practical problems. And once such "new information" is integrated into the wealth of knowledge this can lead humanity to the opening of new paths of research.[55]

Every discovery and invention is a step that could widen our knowledge horizons, each new observation and each new technique, even if they remain unused for a long time, belong to the growing social legacy that Leonardo da Vinci called "second nature."[56] The *notitie nuove* is the lifeblood of the process by which man penetrates the secrets of nature, thus creating a substrate of knowledge that stimulates minds to look for new inventions. It would certainly be excessive and inappropriate to evoke the positivist image of a Biringuccio paladin of the experimental method; however, the epistemological instances placed in *Pirotechnia* do grant the Siennese engineer a place in the tradition of technical knowledge in the late fifteenth and early sixteenth centuries.[57]

Bibliography

Abbri, Ferdinando. 2009. Lawrence M. Principe (ed.), Chymists and Chymistry: Studies in the History of Alchemy and Early Modern Chemistry. *Minerva* 47: 115–118.

Agricola, Georg. 1556. *De re metallica*. Basel: Froben.

Agricola, Georg. [1556] 1950. *De re metallica*, eds. and trans. Herbert Clark Hoover and Lou Henry Hoover. New York: Dover Publications Inc.

Anonymous. [1520 (?)] 1949. Probierbüchlein. In *Bergwerk- und Probierbüchlein*, eds. and trans. Anneliese Grünhaldt Sisco and Cyril Stanley Smith. New York: The American Institute of Mining and Metallurgical Engineers.

---

[54] Biringuccio, *Pirotechnia* (1942), 28. For the Italian, see Biringuccio, *De la pirotechnia* (1977), fol. 1v.

[55] See Bernardoni, *La conoscenza del fare*, 55–63; and Rossi, *I filosofi e le macchine*, 49–52.

[56] See Kemp, "'Wrought by No Artist's Hand'"; Long, *Objects of Art/Objects of Nature*, 63–82; and Findlen, *Inventing Nature*, 297–323.

[57] See, Mieli, "Vannoccio Biringuccio"; Guareschi, "Vannoccio Biringuccio", 432–8; Galluzzi, *Mechanical Marvels*, 11–18; and Smith, *Body of the Artisan*, 142–9.

Arnaud, E.R. 1655. *Introduction a la chymie, ou a la vraye physique, où le lecteur treuvera la definition de toutes les operations de la chymie, etc*. Lyon: Claude Post.

Avicenna. 1927. *De congelatione et conglutinatione lapidum, being sections of the Kitab al-Shifa/ the Latin and Arabic texts*, eds. and trans. Eric John Holmyard and Desmond Christopher Mandeville. Paris: Paul Geuthner.

Beretta, Marco. 2010. Leonardo and Lucretius. In *Rinascimento*, ed. Leo S. Olschki, Series 2, Vol. XLIX, 341–372.

Bernardoni, Andrea. 2007. La quaestio de alchimia nel *De la pirotechnia* di Vannoccio Biringuccio. In *Atti del XII convegno nazionale di storia e fondamenti della chimica*, ed. Franco Calascibetta and Luigi Cerruti, 261–276. Rome: Accademia dei LX.

Bernardoni, Andrea. 2009. Leonardo and the Equestrian Monument for Francesco Sforza: The Story of an Unrealized Monumental Sculpture. In *Leonardo da Vinci and the Art of Sculpture*, ed. Gary M. Radke, 95–135. Atlanta: High Museum of Art, Yale University Press.

Bernardoni, Andrea. 2011a. Elementi, sostanze naturali, atomi: osservazioni sulla struttura della materia nel *Codice Arundel* di Leonardo. In *Il* Codice Arundel *di Leonardo: ricerche e prospettive*, Atti del Convegno Università degli Studi di Bergamo, eds. Andrea Bernardoni and Giuseppe Fornari, 77–105. Poggio a Caiano: CB edizioni.

Bernardoni, Andrea. 2011b. *La conoscenza del fare, ingegneria arte e scienza nel* De la pirotechnia *di Vannoccio Biringuccio*. Rome: L'Erma di Bretschneider.

Bernardoni, Andrea. 2012. Leonardo and the 'Chemical Arts'. *Nuncius* 27: 11–55.

Bernardoni, Andrea. 2013. *Esplosioni, fusioni e trasmutazioni: il monumento a Francesco Sforza e le arti chimiche in Leonardo: disegni di Leonardo dal Codice Atlantico*. Novara: De Agostini.

Biringuccio, Vannoccio. [1540] 1914. *De la pirotechnia*, ed. Aldo Mieli. Bari: Società tipografica editrice Barese.

Biringuccio, Vannoccio. [1540] 1942. *The Pirotechnia of Vannoccio Biriniguccio: The Classic Sixteenth-Century Treatise on Metals and Metallurgy*, eds. and trans. Cyril Stanley Smith and Marta Teach Gnudi. New York: Dover Publications.

Biringuccio, Vannoccio. [1540] 1977. *De la pirotechnia*, ed. Adriano Carugo. Milan: Il Polifilo.

Boni, Bruno. 1954. *Leonardo da Vinci e l'alchimia*. Milan: Societa Editrice la Chimica.

Brioist, Pascal. 2013. *Léonard de Vinci, homme de guerre*. Paris: Alma Editeur.

Brugnoli, Maria Vittoria. 1954. Documenti, notizie e ipotesi sulla scultura di Leonardo. In *Leonardo, saggi e ricerche*, ed. Achille Marazza, 359–389. Rome: Istituto poligrafico dello Stato.

Calvesi, Maurizio. 1986. *Arte e alchimia*. Florence: Giunti.

Cardano, Girolamo. 1539. *Practica arithmetica et misurandi singularis*. Milan: Bernardini Calusci.

Cardano, Girolamo. 1558. *De rerum varietate*. Avignon: M. Vicentinum.

Cattaneo, Girolamo. 1572. *Dell'arte del misurare*. Brescia: Marchetti.

Cipriani, Curzio. 1995. *De re metallica* e *De la pirotechnia*: Agricola e Biringuccio. In *Atti e memorie dell'Accademia Toscana di Scienze e Lettere "La Colombaria"* 46: 9–30.

Conticelli, Valentina. 2012. *L'alchimia e le arti: la Fonderia degli Uffizi, da laboratorio a stanza delle meraviglie*. Livorno: Sillabe.

Di Pasquale Salvatore. 1987. Leonardo, Brunelleschi and the Machinery of the Construction Site. In *Leonardo da Vinci Engineer and Architect*, ed. Paolo Galluzzi, 163–181. Montreal: Montreal Museum of Fine Arts.

Emerton, Norma E. 1984. *The Scientific Reinterpretation of Form*. Ithaca/London: Cornell University Press.

Ferino-Pagden, Sylvia, Francesca del Torre Scheuch, Elisabetta Fedda, and Mino Gabriele. 2003. *Parmigianino e la pratica dell'alchemica*. Milan: Silvana editoriale.

Filarete, Antonio Averlino detto il. 1972. *Trattato di architettura*, eds. A.M. Finoli and L. Grassi. Milan: Il Polifilo.

Findlen, Paula. 2002. Inventing Nature: Commerce, Art and Science in the Early Modern Cabinet of Curiosities. In *Merchants and Marvels: Commerce, Science and Art in Early Modern Europe*, eds. Pamela H. Smith and Paula Findlen, 297–323. New York: Routledge.

Frangioni, L. 1992. *Milano e le sue misure, appunti di metrologia lombarda fra Tre e Quattrocento*. Napoli: Edizioni Scientifiche Italiane.

Frosini, Fabio. 1997. Pittura come filosofia: note su 'spirito' e 'spirituale'. *Achademia Leonardi Vinci. Journal of Leonardo Studies & Bibliography of Vinciana* 10: 35–59.

Frosini, Fabio. 2011. Il concetto di forza in Leonardo da Vinci. In *Il Codex Arundel di Leonardo: ricerche e prospettive*, eds. Andrea Bernardoni and Giuseppe Fornari, 115–128. CB Edizioni: Poggio a Caiano.

Gabriele, Mino. 1988. *Le incisioni alchemico-metallurgiche di Domenico Beccafumi*. Florence: Giulio Giannini and Figlio.

Galison, Peter. 1999. Trading Zone: Coordinating Action and Belief. In *The Science Studies Reader*, ed. Mario Biagioli, 137–160. New York: Routledge.

Galluzzi, Paolo. 1993. Portraits of Machines in Fifteenth-Century Siena. In *Non-Verbal Communication in Science Prior to 1900*, ed. Renato Mazzolini, 53–90. Florence: Olschki.

Galluzzi, Paolo. 1997. *Mechanical Marvels: Invention in the Age of Leonardo*. Florence: Giunti.

Galluzzi, Paolo. 2008. Leonardo da Vinci's Concept of "Nature": "More Cruel Stepmother than Mother". In *Aurora Torealis*, eds. M. Beretta, K. Grandin, and S. Lindqvist, 13–29. Sagamore Beach: Science History Publications.

Garzoni, Tommaso. 1996. *La piazza universale di tutte le professioni del mondo*, eds. Paolo Cherchi and Beatrice Collina. Turin: Einaudi.

Gaurico, Pomponio. [1504] 1999. *De sculptura*, ed. Paolo Cutolo. Naples: Edizioni Scientifiche Italiane.

Gombrich, Ernst H. 1986. Leonardo and the Magicians: Polemics and Rivalry. In *New Light on Old Masters. Studies in the Art of the Renaissance IV*, ed. Ernst H. Gombrich, 61–88. Oxford: Phaidon.

Grellard, C., and A. Robert. 2009. *Atomism in Late Medieval Philosophy and Theology*. Leiden/ Boston: Brill.

Guareschi, Icilio. 1904. Vannoccio Biringuccio e la chimica tecnica. In *Supplemento annuale alla enciclopedia di chimica scientifica e industriale colle applicazioni all'agricoltura ed industrie agronomiche*, ed. Icilio Guareschi, 419–448. Turin: UTET.

Halleux, Robert. 1986. L'Alchimiste et l'essayeur. In *Die Alchemie in der europäischen Kultur- und Wissenschaftsgeschichte*, ed. Christoph Meinel, 187–211. Wiesbaden: Harrassowitz.

Halleux, Robert. 2009. *Le savoir de la main: savants et artisans dans l'Europe pré-industrielle*. Paris: Armand Colin.

Hannaway, Owen. 1986. Laboratory Design and the Aim of Science: Andreas Libavius versus Tycho Brahe. *Isis* 77: 585–610.

Hooykaas, R. 1953. La théorie corpusculaire de Léonard de Vinci. In *Léonard de Vinci et l'expé rience scientifique au XVIe siècle*. Colloques internationaux du Centre National de la Recherche Scientifique, 163–170. Paris: Presses Universitaires de France.

Kemp, Martin. 1995. 'Wrought by No Artist's Hand': The Natural, the Artificial, the Exotic, and the Scientific in Some Artifacts from the Renaissance. In *Reframing the Renaissance: Visual Culture in Europe and Latin America, 1450–1650*, ed. Claire Farago, 177–196. New Haven: Yale University Press.

Kemp, Martin. 2006. *Leonardo da Vinci: The Marvellous Works of Nature and Man*. Oxford: Oxford University Press.

Kiang, Dawson. 1997. Leonardo and Alchemy: A Bibliographical Note. *Achademia Leonardi Vinci. Journal of Leonardo Studies & Bibliography of Vinciana* 10: 199–201.

Lennep, Jacques van. 1985. *Alchimie: contribution à l'histoire de l'art alchimique*. Brussels: Crédit Communal de Belgique.

Leonardo da Vinci. 1973–1980. *Codice Atlantico*, ed. Augusto Marinoni, 24 vols. Florence: Giunti.

Leonardo da Vinci. 1978–1980. *Corpus of the Anatomical Studies in the Collection of Her Majesty the Queen at Windsor Castle*, eds. Kenneth D. Keele and Carlo Pedretti, 3 vols. London: Johnson Reprint Company.

Leonardo da Vinci. 1980–1985. *Corpus degli studi anatomici nella collezione di Sua Maestà la Regina Elisabetta II nel Castello di Windsor*, eds. Kenneth D. Keele and Carlo Pedretti, 3 vols. Florence: Giunti Barbera.

Leonardo da Vinci. 1998. *Il codice Arundel 263 nella British Library*, ed. Carlo Pedretti, trans. Carlo Vecce. Florence: Giunti.

Leonardo da Vinci. 2001. *The Manuscripts of Leonardo da Vinci in the Institut de France: Manuscript L*, ed. and trans. John Venerella. Milan: Castello sforzesco.

Leonardo da Vinci. 2002a. *The Manuscripts of Leonardo da Vinci in the Institut de France: Manuscript E*, ed. and trans. John Venerella. Milan: Castello sforzesco.

Leonardo da Vinci. 2002b. *The Manuscripts of Leonardo da Vinci in the Institut de France: Manuscript F*, ed. and trans. John Venerella. Milan: Castello sforzesco.

Leonardo da Vinci. 2003. *The Manuscripts of Leonardo da Vinci in the Institut de France: Manuscript B*, ed. and trans. John Venerella. Milan: Ente raccolta vinciana.

Long, Pamela O. 2001. *Openness, Secrecy, Authorship: Technical Arts and the Culture of Knowledge from Antiquity to the Renaissance*. Baltimore: Johns Hopkins University Press.

Long, Pamela O. 2002. Objects of Art/Objects of Nature: Visual Representation and the Investigation of Nature. In *Merchants and Marvels: Commerce, Science and Art in Early Modern Europe*, eds. Pamela H. Smith and Paula Findlen, 63–82. New York: Routledge.

Long, Pamela O. 2011. *Artisan/Practitioners and the Rise of the New Sciences, 1400–1600*. Corvallis: Oregon State University Press.

Maccagni, Carlo. 1993. Leggere, scrivere e disegnare la "scienza volgare" nel Rinascimento. *Annali della Scuola Normale Superiore di Pisa*, serie III 23(2): 631–676.

Marinoni, Augusto. 1960. "L'Essere del nulla." *I Lettura Vinciana. Vinci, Bibliotheco Leonardiana*, 26 April 1960. Florence: Barbèra.

Martinón-Torres, Marcos. 2007. The Tools of the Chymist. Archeological and Scientific Analyses of Early Modern Laboratories. In *Chymists and Chymistry: Studies in the History of Alchemy and Early Modern Chemistry*, ed. Lawrence M. Principe, 149–163. Sagamore Beach: Science History Publications.

Michelangelo Buonarroti. 1965. *Il carteggio di Michelangelo*, eds. Giovanni Poggi, Paola Barocchi, and Renzo Ristori. Florence: Sansoni.

Mieli, Aldo. 1914. Vannoccio Biringuccio ed il metodo sperimentale. *Isis* 2: 91–99.

Murdoch, J.E. 2001. The Medieval and Renaissance Tradition of Minima Naturalia. In *Late Medieval and Early Modern Corpuscular Matter Theories*, eds. C. Lüthy, J.E. Murdoch, and W.R. Newman, 91–131. Leiden: Brill.

Nanni, Romano. 2011. Lucrezio: "un ennesimo candidato per la 'filosofia' di Leonardo". *Giornale critico della filosofia italiana* 3: 463–491.

Newman, William R. 1989. Technology and Alchemical Debate in the Late Middle Ages. *Isis* 80: 423–445.

Newman, William R. 2004. *Promethean Ambitions: Alchemy and the Quest to Perfect Nature*. Chicago: University of Chicago Press.

Newman, William R., and Lawrence M. Principe. 1998. Alchemy vs. Chemistry: The Etymological Origins of a Historiographic Mistake. *Early Science and Medicine* 3: 32–65.

Nummedal, Tara E. 2007. *Alchemy and Authority in the Holy Roman Empire*. Chicago: University of Chicago Press.

Partington, James Riddick. 1961. *A History of Chemistry*, vol. 2. London: Macmillan.

Pedretti, Carlo. 1978. *Leonardo architetto*. Milan: Electa.

Perifano, Alfredo. 1996. L'Alchimie dans de la pirotechnia de Vannoccio Biringuccio. *Revue des é tudes italiennes* 42: 189–202.

Perifano, Alfredo. 1997. *L'Alchimie à la Cour de Come Ier de Médicis: savoirs, culture et politique*. Paris: Honoré Champion.

Principe, Lawrence M., and Lloyd DeWitt. 2002. *Transmutations: Alchemy in Art*. Philadelphia: Chemical Heritage Foundation.
Reti, Ladislao. 1952a. Le arti chimiche di Leonardo da Vinci. *La chimica e l'industria* 34(7): 721–743.
Reti, Ladislao. 1952b. Le arti chimiche di Leonardo da Vinci. *La chimica e l'industria* 34(11): 655–667.
Rossi, Paolo. 1971. *I filosofi e le machine 1400–1700*. Milan: Feltrinelli.
Scaglia, Giustina. 1981. Alle origini degli studi tecnologici di Leonardo. XX Lettura Vinciana. Florence: Barbera Editore.
Schneider, Wolfgang. 1949. *Das chemische Wissen Leonardo da Vincis und seine Bedeutung für die Geschichte der Chemie*, Die Pharmazie 4.
Smith, Pamela H. 2004. *The Body of the Artisan: Art and Experience in the Scientific Revolution*. Chicago: University of Chicago Press.
Taylor, Frank Sherwood. 1953. Léonard de Vinci et la chimie de son temps. In *Léonard de Vinci et l'expérience scientifique au XVIe siècle*, ed. Centre National de la Recherche Scientifique, 151–161. Paris: Presses Universitaires de France.
Varchi, Benedetto. 1827. *Questione sull'alchimia: codice inedito*, ed. Domenico Alessandro Moreni. Florence: Stamperia Magheri.
Vasoli, Cesare. 1982. Note su Leonardo e l'alchimia. In *Leonardo e l'età della ragione*, ed. Enrico Bellone and Paolo Rossi. Milan: Scientia.
Zilsel, Edgar. 2000. *The Social Origins of Modern Science*, eds. Diederick Raven, Wolfgang Krohn, and Robert S. Cohen. Dordrecht/Boston/London: Kluwer Academic Publishers.

Artificial Interventions in the Natural Form of Things: Shared Metallogenetical Concepts of Goldsmiths and Alchemists

Henrike Haug

Abstract A mounted coconut, reworked and ornamented by a goldsmith from the *Erzgebirge* and filled with a mountainous landscape, stands in the centre of the present investigation. Its interior is an artificial world assembled from different ores where the anonymous artist depicted Adam and Eve at the moment of the Fall, surrounded by contemporary miners hewing stone and digging for metal using various procedures. A typical *Kunstkammer* object of the sixteenth century, in which the appreciation of rare, precious materials is joined by delight in human virtuosity and invention, the coconut translates montanistic topics and metallurgical lore into material form. This achievement is discussed within the context of natural philosophical theories concerning the causes and the creation of metal in early modern times with the aim of reconstructing the common *Diskursrahmen* and *Vorstellungshorizont*, in which mining professionals, as well as alchemists and goldsmiths worked and in which they localized their practices and self-conception. The mining city St. Joachimsthal in the *Erzgebirge* is introduced as a space where practitioners and scholars worked side by side. Here natural philosophical conceptions were mediated to artisans and craftsmen and empirical know-how and knowledge of materials was conveyed to academics.

How much natural philosophical knowledge and how many alchemical concepts are encapsulated in a sixteenth-century *Kunstkammer* object? And how did goldsmiths come in contact with these ideas: from whom, mediated through which sources and in which setting? Might the goldsmiths have been involved in the development and dissemination of these theories? When goldsmiths loaded an object with references to natural philosophical discourse what was the motivation, the context, and who was the audience?

H. Haug (✉)
Department of Art History and Historical Urban Studies, Technische Universität Berlin, Berlin, Germany
e-mail: henrike.haug@tu-berlin.de

S. Dupré (ed.), *Laboratories of Art*, Archimedes 37,
DOI 10.1007/978-3-319-05065-2_4, © Springer International Publishing Switzerland 2014

The first part of this article attempts to answer these questions in relation to metallogenetic ideas by introducing diverse conceptions of the formation and origin of ore deposits.[1] These contemporary explanations of the creation of metals help explain why knowledge of metals and minerals could affect the working process of both the alchemist and the goldsmith. The goldsmiths' craft knowledge—as well as the information gathered by professionals in the mining districts of the *Erzgebirge* (Ore Mountains), Harz and other regions through the methods of observation, of collecting and of categorization—was incorporated into scientific treatises on the formation of metals and helped to change and develop this domain. The artisan's craft knowledge about matter and materials rarely found its way into a written form. It was most commonly expressed in its suitable mode, the accomplished work of art.[2]

The second part of the article analyses the materials and shape of a coconut vessel on display in the Kunsthistorisches Museum in Vienna (Fig. 1) to see if the aforementioned metallogenetical theories could have had an impact on the form of a sixteenth-century art object: in this *Kunstkammer* object an unknown goldsmith from the *Erzgebirge* created a mountainous landscape, simultaneously a representation of paradise with Adam and Eve eating from the tree of knowledge and the depiction of a post-expulsion mining area populated by contemporary miners. This study suggests some of the possible messages the anonymous goldsmith formulated for the owner and beholder, who was presumably interested in obtaining objects with natural philosophical content for his collection.

A mediator between the metallogenetical research and the goldsmith's *artificium* is present in the person of the St. Joachimthaler preacher, historiographer, and minerals collector Johannes Mathesius (1504–1565). He was an eminent figure in the circulation of alchemical and metallurgical knowledge; through his work it can be seen how this specific codified knowledge was applied and how it could become a common good in montanistic centres and mining areas.

Metallogenetical Concepts

Many theories on the formation of metal were exhaustively compiled in Mathesius's "Die Dritte Predigt. Von Ursprung zu und abnemen der metallen unnd Minerischen bergkarten und Ertzen" (Third Sermon. On the Origin, Growing and Reduction of Metals and Minerals and Ores) (Hereafter Third Sermon).[3] Mathesius published his sermon collection *Sarepta oder Bergpostil* in Nuremberg in 1562. In these texts he incorporated all his accumulated empirical knowledge about mining and minerals (based on his famous collection) enriched by his vast

[1] Adams, "Origin"; and Suhing, "Philosophisches."

[2] For the concept of 'craft knowledge', see Smith, *Body of the Artisan*.

[3] For Mathesius, see Dufek, "Mathesius"; and Loesche, "Mathesius."

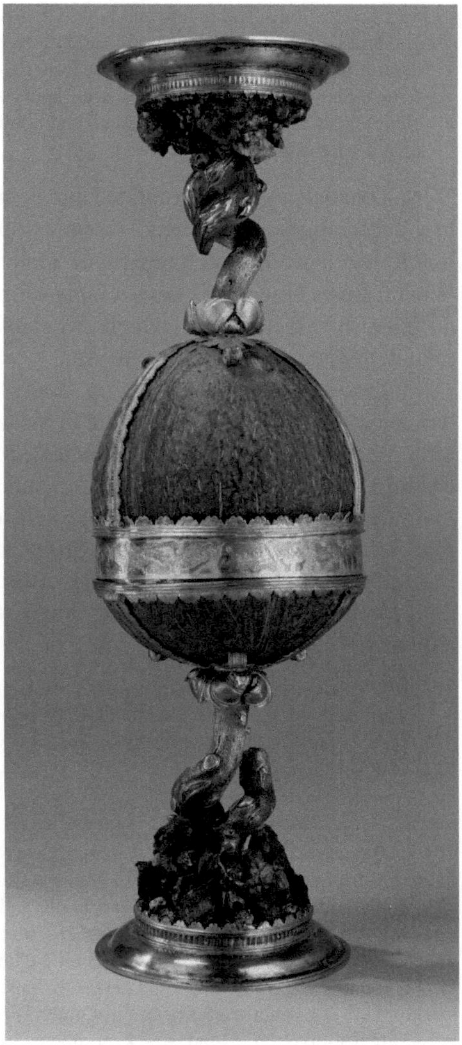

Fig. 1 *Kokosnuss-Doppelscheuer* (*coconut vessel*), closed, sixteenth century St. Joachimsthal (?). Vienna, Kunsthistorisches Inv. No. 885/886

theological learning. The sermons provide an excellent synopsis of the montanistic mindscapes in the early modern period. In the "Third Sermon", held in 1555, Mathesius discusses the emergence of ore deposits and combines several, to some extent competing, early modern metallogenetical theories in one narration for his parishioners, that is, mining professionals from the famous city of the bohemian *Erzgebirge*, St. Joachimsthal.[4] Mathesius preaches:

> [...] about your metals or, as you call them, about certain *Bergarten* and ores and from their names, nature, and characters and how and through and from what they were created in

[4] Mathesius, *Sarepta,* fol. 38r–57r; Fischer, *Bergmännisches* (1969), and *Bergmännisches* (1970).

> their seams, lodes and layers and how one metal can be transformed into another until the good metal [is achieved], massive and completely. And how the spare metal again through natural heat is reduced and consumed, as the holy scripture and honest and experienced miners testify and I myself have seen and learned and is tolerable for and suited in the church, and I will give report in plain and good montanistic words, to identify and praise in it God's hand and creation.[5]

Mathesius's clearly formulated aim is the recognition of God in his opus (natural theology), more specifically, as he is speaking to the mining professionals of the *Erzgebirge*, the recognition of God in one of his marvellous gifts: metal. What follows however is not a simple sermon appropriate for the church, but rather a scientific lecture in which Mathesius presents and discusses different metallogenetical positions of his time.[6]

Mathesius starts by comparing God's formative acts with the practice of the human artisan, who imitates nature in his workshop with the substances available to him, stating "God has various *schmeltzwerck* [products produced by melting/fusion/casting] in his *laboratorio* [workshop] and smelts the metals as beautifully and in as many colours as the flowers in the field or the crests of a stonecutter."[7] God is personified as some kind of metalworker, maybe a goldsmith in his workshop, who uses a smelting process to form the several metals that can be found in the earth. And Mathesius doesn't leave any doubts that this is a perpetual process when he explains:

> The experience of every day shows—aside from scholarly reports—that our God daily creates and lets grow in the earth all kinds of noble and rough stones as well as the various precious and salubrious *bergsafften* [juices of the mountain] and various *bergarten* [minerals], ores and pure metals.[8]

---

[5] "von ewern metallen oder wi ir pflegt zu reden von allerley bergarten unnd ertzen unnd von iren namen, natur unnd eygenschafft und wie und warauß und wardurch sie in iren gengen, fletzen und stöcken gewircket unnd wie ein bergart in die ander verwandheit biß die gültigen metal, gedigen und vollstendig und wie die uberstendigen metal wider durch natürliche hitz inn der erden abnemen unnd verzert werden, so viel mich die heylige schrifft und ehrliche und erfarne bergleut berichtet und ich selbs gesehen und gemerckt habe und sich in der kirchen leyden unnd gezimmen wil unnd diß alles euch einfeltig und mit guten bergkleufftigen worten berichten, doch das hierinn vornemlich Gottes hand und werck gerühmet und erkannt werde." ("Die Dritte Predigt. Von Ursprung zu und abnemen der Metallen unnd Minerischen Bergkarten und Ertzen," in Mathesius, *Sarepta*, fol. XXXIIIv).

[6] The concept of nature as 'second revelation' is connected with the metaphor of 'reading' in the book of nature, see Blumenberg, *Lesbarkeit*; Nobis, *Buch der Natur*; Rothacker, *Buch der Natur*; and Dülmen, *Buch der Natur,* 131–50.

[7] "Denn Gott hat mancherley schmeltzwerck inn seinem laboratorio und schmeltzet je die metal so schön und vilerely farben als die blümlein auff dem felde oder ein Steinschneider sein wapenstein." ("Von Ursprung zu und abnemen der metallen unnd Minerischen bergkarten und Ertzen," in Mathesius, *Sarepta*, fol. XLIr).

[8] "Es gibet teglicher erfahrung neben der gelerten zeugnuß das unser Gott inn der erden allerley edel und gemeine steyne neben mancherley köstlichen unnd heylsamen bergsafften unnd mancherley bergarten, ertzen und reinen metallen teglich schaffe und wachsen lasse [. . .]"("Die Dritte Predigt. Von Ursprung zu und abnemen der Metallen unnd Minerischen Bergkarten und Ertzen," in Mathesius, *Sarepta*, fol. XXXIXr).

Mathesius, in line with most contemporaneous authors, assumes a continuing emergence of substances with God, or a divine potency incorporated in nature, responsible for the on-going generation of metals.[9]

The main issues that concerned Mathesius and contemporary writers were: what substances are the sources of the metals? Are there one or more primary matters? How and under what conditions do these primordial materials transmute into the different metallic substances? And last but not least: what causes this process?

Mathesius and his "Third Sermon" is not a randomly chosen source. The author stood in the centre of scholarly culture in St. Joachimsthal, in one of the most important centres of the Renaissance mining industry and he was an important link in the vibrant network of many protagonists involved in mineralogical and metallurgical research. Mathesius was a friend and interlocutor of Georg Agricola (1494–1555), one of the eminent representatives of early modern metallurgy, and was familiar with his writings as well as with older German metallurgical literature. He read both Ulrich Rülein of Calw's (1465–1523) *Bergbüchlein* published in 1505 and the more modern *De la pirotechnia* by Vanoccio Biringuccio (1480–c.1537) from 1540.[10] He was a well-known collector of minerals and well connected in the intellectual world of St. Joachimsthal. Furthermore, as a student and dining partner of Martin Luther (1483–1546) he operated in Protestant humanist circles: in short an authority *par excellence* for all themes concerning montanist mindscapes of the sixteenth century.

Mathesius's texts interweave preaching and scientific imagination and thus attest to how strongly early modern natural philosophy and metallurgic research are influenced by the Christian episteme:

> But my undertaking shall conduct to that effect, to show to you, my parishioners, the almighty and wonderful hand of God and his immeasurable abundance and his inscrutable wisdom and his merciful and fatherly heart in his creation and revelation of all kinds of minerals and metals, to make you recognize your God in his gifts and teach you to praise him, which he conveys to you in this mountain in clement benevolence.[11]

This juxtaposition of empirical knowledge based on experience next to philosophical and theological reflections of the causes of things is typical for Mathesius and many other writers of the sixteenth century, as they interpret their increasing

[9] This concept can be found in medieval authors writing on *natura* as goddess, as *pro-dea* and as *vicaria* (assistant) of God, who was entrusted with the *creatio continua*, the continuous creations as *pro-creatrix*, like Alanus ab Insulis or Jean de Meuns in his roman de la rose. See Modersohn, *Natura als Gottin*.

[10] Biringuccio, *Pirotechnia* [1540], and *Pirotechnia* (1942).

[11] "Mein vornehmen aber sol sich eygentlich dahin lenden, das ich euch meinen Pfarrkindern vornehmlich die allmechtige und wunderbarliche hand Gottes und seinen unmeßlichen reychtumb neben seiner unerforschlichen weyßheyt unnd genedigen unnd Väterlichen hertzen inn Schöpffung und offenbarung allerley ertz unnd metalle zeyge, damit ir ewern Got in seinen gaben erkennen und preysen lernet, die er euch inn diesem gebirge auf genediger güte mittheylet." ("Die Dritte Predigt. Von Ursprung zu und abnemen der Metallen unnd Minerischen Bergkarten und Ertzen," in Mathesius, *Sarepta*, fol. XXXIIIv).

empirical knowledge within the limits of a regular and harmonic world, where everything points back to its origin, i.e. God.

Mathesius and other mineralogists of the sixteenth century could only rely on a few preliminary works when they made their speculations on the *Metallogenese*, as Agricola regrets in the introduction to his *De ortu et causis subterraneorum* from 1546:

> The Greeks and Latins, however, who we see working for more than a thousand years to increase scientific knowledge, all of them only interpreted the writings of Plato and Aristotle and followed their views; they neither made inquiries about unsolved questions nor did they treat them scientifically. When our Albertus started to make observations about the genesis of excavated materials, he blended the teachings of philosophers, astrologers and chymists into one.[12]

Two important preliminary writings, Aristotle's reflections on meteorology and Albertus Magnus's (1193–1280) *Libri cinque de mineralibus et rebus metallicis*, are considered by Agricola and other early modern authors and are present in almost all theories of the sixteenth century. With Albertus's *De mineralibus* an important testimony to the transmission of antique metallurgic knowledge and its combination with alchemical concepts has been preserved. That Albertus's knowledge was not out of date in the early modern period is attested by the two prints of his works by Giovanni and Gregorius de Gregoriis in Venice (*De mineralibus*, 1495) and of Jacob Köbel (*Liber mineralium,* 1518) in Oppenheim. His amalgamation had a great influence and contributed strongly to the work of several natural philosophers and alchemists of the sixteenth century. It found its way equally into alchemical practices as well as the theoretical concepts of 'modern' metallurgists and was part of the common knowledge in the workshops of goldsmiths and other metalworkers of the early modern periods. A remark in Aristotle's treatise lead Albertus to assume that the ancient philosopher had written a follow-up work on minerals, stones, earth and metals. It was Albertus's intention to comment on the whole Aristotle, i.e. to commentate the texts that were in the thirteenth century considered to be the Aristotelian philosophical works on nature. Therefore it is no wonder that he, believing that Aristotle's treatise was lost, commenced his own *Lapidary* (Book of Stone): "We have not seen Aristotle's books about these [minerals], but only some excerpts from them; and what Avicenna says about [minerals] in the third chapter of the first book which he wrote about them is not sufficient."[13] Thus, Albertus's text is an advancement of the Aristotelian systems, which he blends with alchemical texts and concepts obtained from Arabic sources. He also enriched his text with empirical know how gained on his visits to mining

[12] "Quos vero tam Graeco, quam Latinos annis plus mille in rerum congitiones versatos videmus, omnes ad unum scripta Platonis, aut Aristoteles interpretati sunt, & sententias fecuti: eoque minus multas res nondum, explanatas vocarunt in quaestionem, literisque illustrarunt. Nam Albertus noster aggressus dicere de ortu eorum quae effodiumtur, Philosophorum & Alstrologorum & Chymistarum decreta in unum confundit." (Agricola, "Geologie und Mineralogie," 76, and *Ortu*, 24).

[13] Albertus, *Book of Minerals*, 9.

districts as he relates in his *Book of Minerals* (III, I, I). According to Aristotle, there are four causes: the material, the efficient, the formal and the final.[14] Albertus begins with the material cause discussing the primary matter from which different minerals were created. Naturally Albertus's chemical understanding is also shaped by Aristotle, so he starts with the four elements: fire, water, air and earth. The general classification throughout his text on minerals is based on this material cause: Albertus starts with a chapter on stones, which he classified as simple structures made from a mixture of earth and water. According to his system, metals are more elaborate structures built from sulphur and mercury (quicksilver), which themselves are mixtures of earth and water (quicksilver) and all four elements (sulphur). This sulphur-quicksilver theory is non-Aristotelian, but was an adaption from Arabic writings.

The second part of Albertus's system is built by the efficient causes—the agents or operators that cause the creation of minerals and metals from a still non-specific primordial matter: Albertus takes the 'two *exhalationes* theory' of Aristotle and adjusts it a little, by equating dry smoke with sulphur and moist vapour with mercury. According to Aristotle there are two kinds of aeroform emanations which ascend from the middle of the earth to the surface. One of these *exhalationes* is a dry fume which transforms into earth and stones, the other a humid steam, which condenses into metals:

> We maintain that there are two exhalations, one vaporous the other smoky, and there correspond two kinds of bodies that originate in the earth, 'fossils' and metals. The heat of the dry exhalation is the cause of all 'fossils'. Such stones cannot be melted [. . .]. The vaporous exhalation is the cause of all metals, those bodies which are either fusible or malleable such as iron, copper, gold. All these originate from the imprisonment of the vaporous exhalation in the earth, and especially in stones. Their dryness compresses it, and it congeals just as dew or hoar-frost does when it has been separated, though in the present case the metals are generated before that segregation occurs. Hence, they are water in a sense, and in a sense not [. . .]. Hence, they all (except gold) are affected by fire, and they possess an admixture of earth; for they still contain the dry exhalation.[15]

To convert these two *exhalationes* into ore deposits, Albertus says that they must: first, be confined in the earth; and second, react under the power of two instruments, that is heat and cold. But the transformation from primordial matter into the several metals is not a mere mechanical process, it requires a mineralizing power to achieve the destined form.

Here Albertus merges an alchemical concept with ideas from Aristotle extracted from his *Generation of Animals*. He equates a female principle with the material cause, i.e. the shapeless matter, subjected to an efficient cause, a forming principle that is inherent in the male semen. To initiate this transformational process Albertus introduces the celestial bodies as impulses. After discussing the material and efficient cause he then treats the third, the formal cause, where he follows Aristotle who said that the male principle contributes the form of the offspring, i.e. the

---

[14] Albertus, *Book of Minerals*, XXXI.

[15] Aristoteles, "Meteorologie," 136.

species. Again Albertus turns to the power of the stars that act as formative powers descending from the heavens and which are responsible for the seven known main metals.

The structure of the book and the line of his argumentation shows that Albertus at one point had to face the question of transmutation: if metals are formed from primordial matter consisting of a mixture of the four elements, i.e. from a mixture of sulphur and mercury as substances in between the basic elements and the finished metals, and if metals are constantly emerging as defined species, then this logically leads to a discussion of whether one metal can be transformed into another as the alchemists claim. Albertus does not totally dismiss the possibility of transmutation, but—as a follower of Aristotle—accepts the theoretical possibility, without showing particular interest in this question, when he states:

> We do not intend here to show how any one of these may be transmuted into another, or how, by the remedy of that medicine the alchemists call the elixir, their diseases may be cured, or their occult properties made manifest, or conversely their manifest properties be removed. But instead we shall show how they are mixed from the elements, and how each one is constituted in its own specific form.[16]

Albertus, with his system and his reception of ancient philosophical considerations, had a strong influence on the early modern period: he describes properties of the metals like colour, malleability, melting point and so forth as accidental properties of the metals. The ancient and medieval search for the *causes* and the *principles* proceeded to dominate early modern metallogenetic theories. The progeny of these Aristotelean-Albertian concepts are found everywhere in Mathesius, who says about silver:

> Massive or compact silver is the name for what is pure and nearly fine and can be cut and embossed, before it is put in the fire. But all massive and particularly all different forms of *weyß und rotgüldig ertz* [silver ores] and *glaß ertz* [silver sulfid] wear off in the fire because of all the sulphur and the quicksilver and the other impurities that are still in it.[17]

The same with gold: "I have seen gold in an iron-stone, as well as gold in the *Styria* [Steiermark], which changes its golden colour in the fire, when the quicksilver exhales from it, comparable to the goldsmith gilding silver dishes."[18] Both quotations show how deeply Mathesius's understanding is based on the two basic

---

[16] Albertus, *Book of Minerals*, 10.

[17] "Gedigen oder derb silber heyst, das rein und schier fein ist und das sich schneyden und pregen lesset ehe es ins fewer kompt. Doch gehet allem gediegen und sonderlich weyß unnd rotgüldig ertz so wol als dem glaß ertz im fewer was abe umb des Schwebels, quecksilbers und ander wildigkeyt willen, so noch drinne ist." ("Die Dritte Predigt. Von Ursprung zu und abnemen der Metallen unnd Minerischen Bergkarten und Ertzen," in Mathesius, *Sarepta*, fol. XLr). *Weyß und rotgüldig ertz* are old-fashioned names for silver ores, like *Proustit* and *Pyrargyrit*, i.e. minerals containing silver.

[18] "Ich habe sichtig gold in einem eystenstein gesehen, wie auch in Steirmarck weiß gold bricht, welches im fewr sein natürliche farbe bekompt, so das quecksilber davon verauchet, als wenn ein Goldschmid die silber geschir vergüldet." ("Die Dritte Predigt. Von Ursprung zu und abnemen der Metallen unnd Minerischen Bergkarten und Ertzen," in Mathesius, *Sarepta*, fol. XLv).

substances, sulphur and mercury, as the alchemical *Urbausteine* (building blocks) of silver and gold, that will be reduced to unwanted components of the 'pure' metal if the ore is re-heated in the fire in the smelting process. Agricola also discusses the divine origins and causes of minerals when he states in *De ortu et causis* (1546):

> The philosophers, astrologers and chymists study the *ortus et causas* [origins and causes] of ores, because nothing happens without a cause, and they state correctly that even now, as before, the formation of ores takes place. The less acceptable is the completely ludicrous view of the large mass—more a fairy tale, that contradicts all experience—claiming that not only rocks, but also the minerals and gems and different kinds of soil in their burrows, chasms and grooves have been formed at the beginning of the world by God, as they are now found, and in the meantime no such things were created or re-created from possible substances, and the highest [=divine] *opifex* [workman] did not set any sort of natural force in motion to ensure their constant generation.[19]

Another important protagonist of early modern mineralogy, the metallurgist, *Münzmeister* (moneyer) and *Probierer* (tester) Lazarus Ercker (1528–1594) says in his *Beschreibung allerfürnemisten mineralischer Ertzt and Berckwercksarten* (Description of the Most Distinguished Ores and Minerals), his *Großes Probierbuch* of 1574, that he will not discuss the writings of natural philosophers on ores and metals, because:

> [their] thoughts and delusions are not only uncertain and often wrong, but frequently contradict themselves [and] I want to simply believe that God, the almighty Creator, has reserved these secrets in his omnipotence and he created gold and silver as well as all other metals through his word, through which also originates heaven and earth and everything that is on it and in it.[20]

Following this argument, Ercker assumes firstly that all metals were once created at the beginning of time by God, that secondly no genuine natural power continues to create ores and that, consequently and thirdly, the generation of metals and minerals must lie outside human comprehension.

In their criticism of impractical, non-empirical 'book-learned' scholars, Ercker and Agricola concur. They also concur in their demarcation of fraudulent alchemists (the charlatans and quacks) from expert alchemists in whose experiential knowledge they trust. Both scholars base their works on the *Erfahrungswissen* of

---

[19] "Sed Philosophi, Chymistae, Astrologi & ortus metallorum causas perquirunt, quia nihil causa fiat: & metalla nunc, ut quondam, ortu generati recte adferunt. Quo minus est ferenda vulgi opinio perridicula, poetearumque alicujus fabulae similis, & omni experimento contraria: quod ait, non modo faxa, sed in eorum venis, fibris, commissuris metalla & gemmas, ac varia terrarum genera in initio mundi Deum finxisset alia & creasse, cujusmodi nun inveniri solent: nec medio emptoris curricolo nulla, ex materiis ad id aptatis, esse nata, vel renata. Neque aliquam vim naturae ad eorum perpetuitatem machinatum ese summus rerum opificem." (Agricola, *Ortu*, 125–6). For an overview of Agricolas's understanding of matter, see Beretta, *Enlightenment of Matter,* ch. III.

[20] "Gedanken und Wahnvorstellungen [sind] nicht allein ungewiss und oftmals falsch, sondern sich auch häufig wiedersprechen [...] ich will einfältig daran glauben, dass Gott, der allmächtige Schöpfer, diese Geheimnisse seiner Allmacht vorbehalten will und dass er Gold, Silber sowie alle Metalle durch ein Wort erschaffen hat, durch das ebenso Himmel und Erde und alles, was darauf und darinnen ist, ihren Ursprung haben." (Ercker, *Erze und Bergwerksarten*, 283).

the practitioners, Ercker out of his professional experience and his contacts as tester of ores and as *Münzmeister*, Agricola from his medical knowledge and his connections with the network of mining specialists surrounding Mathesius. When Agricola worked as a doctor in St. Joachimsthal, among his acquaintance he numbered the *Hüttenschreiber* Lorenz Wermann, the prototype of the learned mining engineer who served as model for Agricolas's first montanistic publication, the *Bermannus* from 1530.[21]

Agricola, like Ercker, criticises scholars who lack practical experience and can only rely on handed down theoretical knowledge, that is, those who have *episteme*, but not *techne*. Then he criticises the *chymists*, i.e. the alchemists, to whom he grants some empirical and practical knowledge, but most of whom he dismisses as artificers of fraud and who are not on the same level as 'real' metallurgists.[22] Thirdly Agricola takes on the astrologers, referring to the authors who claim that the seven planets as formative powers influence some kind of primordial metallic arch-matter to create the seven known metals. The theory that celestial bodies influence and shape earthly substances is treated at the beginning of Rülein of Calw's *Bergbüchlein*:

> For the formation and growth of ores an effector is required and a subordinated material or matter, which is adapted to receive the effect. The common effector of the ores and of all things that are born is heaven with his orbits, his light and his influence, as the masters of nature teach. [. . .] Every ore receives a special influence from the planet he is named after, so that the planet and the ore concur in their warmth, coldness, humidity and aridity.[23]

The early modern efforts to categorize the world and its substances, based on large collections of minerals and the improved material knowledge of the natural scientist, started to challenge these obsolete systems. This is evident in Mathesius's "Third Sermon," where he names the planets and their connection and influence on the seven main metals. He distinguishes gold, golden silver, silver, copper, iron, steel, lead, tin, bismuth, mercury and *spießglas* and thus clearly enumerates more than seven metals. Mathesius's text probably refers to the writings of Agricola, who emphasises the autonomy of bismuth and who uses the addition of this eighth species to the main metals as his most powerful argument against the star-forming theories of metals:

> According to the opinion of the astrologers, the planet is constantly under the influence of forces that work on suitable substances in the womb of the earth and this create the ores. This tale seemed so nice to many people that they said that the forces of the stars operate in

[21] Quasi a prototype of Edgar Zilsels's 'superior craftman,' see Zilsel, "Roots."

[22] For the role and differing evaluations of the 'alchemist' in the sixteenth century, see Nummedal, *Alchemy and Authority*.

[23] "Zur Entstehung und zum Wachsen der Erze gehört ein Wirker und ein unterworfener Stoff oder Materie, geeignet, die Wirkung zu empfangen. Der gemeinsame Wirker des Erzes und aller Dinge, die geboren werden, ist der Himmel mit seinem Lauf, Licht und Einfluss, wie die Meister der Natur lehren. [. . .] Jedes Erz empfängt einen besonderen Einfluss von den Planeten, nach dem es genannt ist, so dass der Planet und das Erz in ihren Eigenschaften, in Wärme, Kälte, Feuchtigkeit und Trockenheit übereinstimmen." (Calw, *Bergbüchlein*, 117).

> the earth, as if they were second planets [...] where every metal has great resemblance to his planet, like offspring have in relation to their father.[24]

He ends with the provocative question: "But since there are only seven planets, which will they call the producer of bismuth?" and concludes that the forming influence of the stars on earthy matter is a futile conceit.

The second issue in discussion (after the issue of the 'cause', i.e. the creator, who shapes the species of the metals) was the question of the substances or materials from which metals were generated, thus of the primordial matter from which the generation can commence.

In the texts three main theories can be distinguished: firstly that of an *ens primum* or a *materia prima*, that is a more or less solid primordial matter, secondly one that proposes a kind of metal juice, called *Ghur* or *nobilis succus* and thirdly the Aristotelian *exhalationes*, rising moist vapours from the centre of the earth. One or more forming causes affect these primordial matters that all have the potentiality to be transformed, so that at one point this matter mutates into one of the known specific metals, and consequently the different ore-deposits were formed. For example, in his *Bergbüchlein* Rülein of Calw mentions mineral vapours, remotely reminiscent of Aristotle and Albertus, that rise as evaporation from the inner earth and condense to form ore deposits in the corridors and chasms of the mountains. He then changes direction and ties these *exhalationes* back to alchemical concepts:

> The subordinated thing, the common matter of all metals is according to the opinion and faith of the scholars sulphur and quicksilver. Several believe, that due to the orbit and influence of the heaven, exhalations and vapour of sulphur and quicksilver—called *exhalationes minerales*—were drawn up and, during their ascent under the influence of the planets, connected in the chasms and gaps and were made into ores.[25]

The whole range of early modern competing ideas of metallogenesis is revealed when Rülein of Calw in the very next sentence introduces a *Ghur* (also discussed in the "Third Sermon" of Mathesius) as a kind of "humid, cold, mucous, completely sulphur-free matter, which is extracted virtually as sweat of the earth" as an equally valid theory, a humid and mucilaginous element able to transmute under the

---

[24] "Astrologorum autem sententia est, errantes stellas influxu & viribus, quas exercent in materia, ad id in terrae visceribus aptata, efficere metella: inerrantes vero gemmas. Quod figmentum pulchrum qibusdam visum est adeo, ut dicerent, vires stellrum in terra efficere metalla, tanquam secundas, quasdam stellas vagas: & gemmas, tanqam secundas quasdam inerrantes stellas: qualibet gemma obtinente, ex eorum sententia maximam virium su fideris partem; & quolibet metallor similitudinem magnam habente, cum so sidere errante ut proles habet ad parentem." (Agricola, *Ortu*, ch. IX).

[25] "Das unterworfene Ding, der gemeinsame Grundstoff aller Metalle sind nach Meinung und Glaube der Gelehrten Schwefel und Quecksilber. Etliche glauben, dass durch den Lauf und Einfluss des Himmels aus der Tiefe der Erde von Schwefel und Quecksilber Ausdünstungen oder Brodem – exhalationes minerales genannt – aufgezogen und während des Aufsteigens in Gängen und Klüften durch die Wirkung der Planeten vereinigt und zu einem Erz gemacht werden." (Calw, *Bergbüchlein*, 117–8).

addition of sulphur into one specific metal.[26] Agricola discusses both the Aristotelian theory of steam or vapour and the alchemical sulphur-mercury theory and rejects both. He states that the Earth's crust absorbs water and earthy matter, thus creating a certain *nobilis succus* as kind of mineral liquid that sediments in the corridors and chasms of the mountains, there forming metallic minerals.

The hitherto discussed sources of early modern writers on metallogenetic questions have demonstrated—despite the abundance of variety in the existing concepts—that one basic assumption is nearly always present. Almost all theories offer a dualistic system that operates on the dichotomy of a primary passive substance acted on by a potent cause. The involved antagonists are matter and an active formative power, or in other words: a creator and a material in which the creation can manifest. This duality of primordial passive matter and active forming principle can be thought of in terms of natural procreational processes which approximates the third—mineralistic—to the other two reigns, the floral and the animalistic, where by seed or semen procreation and growth is initiated. If this biological analogy taken from the animal and herbal kingdoms is applied to the mineral, it can be extended to metallogenesis: if in the field of animals and plants male and female beings can be found who procreate by the union of the active and formative male seeds or semen and the female passive receiving matter—and if this means that this species can recreate self-reliantly— then it is not too far fetched to suspect comparable ways of reproduction and growth in the reign of minerals and assume the existence of metal 'semen'.

One of the advocates of the equation of all three kingdoms of nature, who awards metals the power of self-propagation, is the famous potter and scientist Bernard Palissy (1510–1589) in his *Discours Admirables*.[27] In this text in the form of a dialogue he has *practique* confront *théorique*. The latter represents scholarly and alchemical positions that maintain the possibility of mimicking metallogenetic processes by human art. For *practique* (who represents Palissys's own positions, as he poses as a learned artisan using his craft knowledge) the belief that man can imitate the creation of metals is presumptuous. *Practique* asks *théorique*:

> Do you think that I want to believe a Gerbert or a Arnaud de Villeneuve or a Roman de la Rose in what they say against the works of Gold? And do you think that I am so misinformed, that I do not know very well that Gold and Silver and all the other Metals are a divine work and that it is a temerarious enterprise against the glory of Gold to try to usurp that which is his realm?[28]

---

[26] "Feuchten, kalten, schleimigen, durchaus schwefelfreien Grundstoff, der gewissermaßen als ihr Schweiß aus der Erde gezogen ist." (Calw, *Bergbüchlein*, 118).

[27] On the work of Palissy, see Amico, *Palissy*; Hanschmann, *Palissy*; Kayser, "Intellectual and Artisan"; "Kemp, Philosophical Pots"; Kris, *Stil Rustique*; Laube, "Wissenswelten"; Schmeisser, "Erdgeschichte"; Shell, "Casting Life"; and Klier, *Fixierte Natur*.

[28] "Cuides tu que je vueille croie un Gerbert ou Arnauld de Vileneufve ou un Romant de la Rose, en ce qu'ils auront parlé contre les euvres de Dieu? Et cuides tu que je sois si mal instruit, que je ne sache bien que l'or & l'argent & tous autres metaux sont une euvre divine, & que c'est temerairement entrepris contre la gloire de Dieu, de vouloir usuper sur ce qui est de son estat." (Palissy, "Discours admirables," 105–6).

Hereupon *theorique* asked directly after the analogy of semen in plants and metals.[29] Palissy/*practique* then argues against the alchemists who claim to imitate in their *vaisseaux*, which "serve as a womb for the generation of metals," the natural creation and "that they want to imitate the uterus of a woman or an animal."[30] According to Palissy this is impossible, because "the matters of the metals are divine seeds. I say on a level divine so that they are unknown to men: even invisible."[31]

For Palissy the metal seed is a real fact but also an analogy for the alchemist who counterfeits God's own work of growing metals, a divine process, which will remain encoded from man, because man (i.e. the scientist or the goldsmith) can only work with the materials provided by God:

> It is not without reason that I have said it is a work of God and that the seeds are the matter of metals and that He gives them growth and to men [the ability] to collect and to purify and to test, to melt and to ally, to cast them [the metals] into some forms as men see fit and useful to them.[32]

Agricola, however, rejects vehemently the idea of a semen-like origin of the metals in *De Ortu et Causis*: "[...] none of the things that occur in the earth can produce something similar to itself [...] The minerals are not provided with seeds that are capable of producing."

The Doppelscheuer

Against the background of these briefly outlined metallogenetic concepts I will now discuss the *Doppelscheuer* (coconut vessel), from Vienna (Fig. 1).[33] When opened, the interior of the coconut reveals a very interesting scene: an artificial mountainous landscape in which Adam and Eve are placed next to the tree of knowledge, all assembled by the hand of a goldsmith using different kinds of ores and semi-precious stones (Fig. 2). This object is a variation of a *Handstein*—an exceptionally large or beautiful mineral that was sorted out of the smelting process and integrated in the context of the *Kunstkammer*, either in its natural form or transformed and

[29] "tu parles icy de semer; comme si les metaux venoyent de semence, comme le bled ou autres vegetatifs." (Palissy, "Discours admirables," 106).

[30] "servir comme une matrice à la generation des metaux."; "voulant imiter la matrice de la femme ou de la beste." (Palissy, "Discours admirables," 107).

[31] "les matieres des metaux sont semence divines. Je di tellement divines qu'elles sont inconnues aux homme: voire invisible." (Palissy, "Discours admirables," 108).

[32] "Ce n'est pas sans cause que je'ay dit que c'est l'euvre de Dieu que de semer la matiere des metaux & leur donner l'accroissement, & aux hommes de les recueillir, purifier & examiner, fondre & mallier, pour les mettre en telle forme que bon leur semblera, pour leur service." (Palissy, "Discours admirables," 108).

[33] Doppelscheuer, Ores into a Coconut, Kunsthistorischen Museum in Vienna, Inv. No. 885/886; Storczer, *Handsteinsammlung*, cat. no. 1: 68–71.

Fig. 2 *Kokosnuss-Doppelscheuer* (*coconut vessel*), open showing both halves, sixteenth century St. Joachimsthal (?). Vienna, Kunsthistorisches Inv. No. 885/886

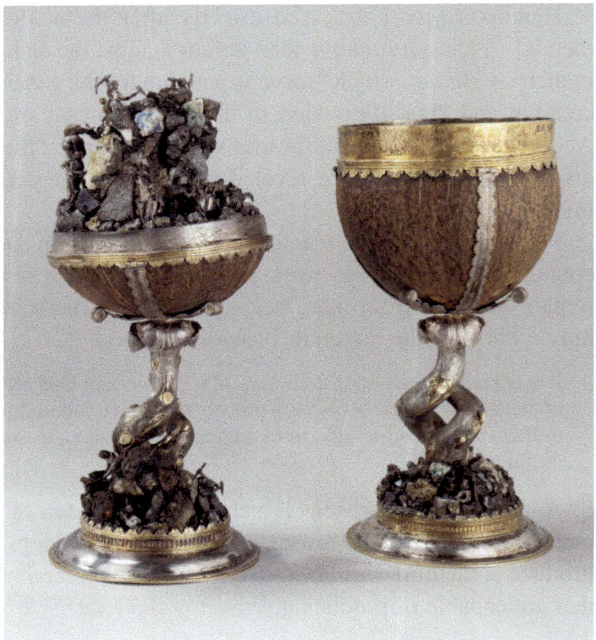

reworked by an adept goldsmith.[34] Both the material and content of the piece suggest it originated in the *Erzgebirge*, the most important mining region of early modern Europe. *Handsteine* are mainly found in the collections of the Saxon Elector August (1526–1586) in Dresden and in *Schloss Ambras*, as its owner, the Archduke of Tyrol, Ferdinand II (1529–1595) as governor of Bohemia also held shares in the Ore Mountains.

The craftsmanship of the local goldsmiths, especially in St. Joachimsthal, with their specialist treatment of this type of material, is testified by Mathesius, who praises in his "Tenth Wedding Sermon" that he saw

> [. . .] thank God in this valley [St. Joachimsthal] a lot of beautiful histories [depiction of stories] from the Old and New Testament, and from respectable and decent pagan stories, stamped unto *Schaugroschen* [coins not coined for money transactions, but to be exhibited] and cut into ores [. . .] I could name a lot of beautiful coins and ores [*Handsteine*], prepared here in this valley, in which—besides felicitous craftsmanship—much of the beautiful contents of the true religion can be seen.[35]

[34] Haug, "Gewechse"; Schlosser, *Wunderkammern,* 50–1; Schiedlausky, "Handsteine"; Strieder, "Erzstufe"; Quellmalz, "Materialfrage"; Distelberger, "Gold und Silber"; *Meisterwerke,* cat. no. 244 a–k: 562–88; Huber, "Stuffe"; *Bei diesem Schein,* 122–35; and Dupré & Korey, "Kunstkammer."

[35] "Gott lob in diesem thal viel schöner Historien, auß altem vnd newem Testament, auch auß erbarn und züchtigen Heydnischen Historien, auff schawgroschen gepreget und in ertz geschnitten sind [. . .] Ich köndte viel schöner groschen und stuffen erwehnen, die hie im Tal zugericht, darinn neben trefflicher kunst, viel schöner artickel der wahren Religion zu sehen sindt [. . .]" (Mathesius, "Hochzeitspredigten," 186).

In 1577 Emperor Rudolph II instructed the Bohemian Chamber:

> We graciously inform you, that in St. Joachimsthal [lives] a goldsmith called Caspar Ulich; he has about sixteen pieces of 'red silver ore', which we are entitled to. And because we would like to have them, we graciously command you to instruct in our place our *Münzmeister* in St. Joachimsthal, to request these pieces of minerals from the goldsmith in the near future, and instruct him to send them to the Bohemian Chamber packaged in such a way, that they will not take damage and that you—when they have arrived—will send them to us immediately.[36]

This letter testifies that Caspar Ulich must have been one of the leading goldsmiths in St. Joachimsthal and was associated with the artistic transformation of valuable minerals—and that the Emperor was much aware of him and controlled very carefully what materials were sent to his workshop.

A second source attests further that St. Joachimsthal must have had a monopoly on the knowledge of processing these *Handsteine*: the widow of the goldsmith Ruprecht Puellacher, who served as mintmaster of St. Joachimsthal, offered in 1564 a *Handstein* reworked by her husband to Emperor Maximilian II for the sum of 7000 Taler.[37] The emperor asked his brother, the Archduke Ferdinand II, as a connoisseur of such works, to estimate the value of the hand stone. Ferdinand replied to his brother in a letter dated 12 July:

> The estimate is wanting, as here in Bohemia there are no similar goldsmiths or artists who understand how to treat such materials, and if the goldsmiths of St. Joachimsthal who worked on such *Handsteine* for several years were to be asked, it is to be feared, that they may well esteem and respect this artwork so highly that it will be very expensive for your imperial Majesty.[38]

Ferdinand admits that the estimation of these objects is not easy due to the fact that in Bohemia only the goldsmiths of St. Joachimsthal are specialized enough in the transformation of minerals into *Handsteine* to make a valuation and thus the possibilities of comparison are lacking.

---

[36] "Wir fuegen euch genedigist zu wissen, das im Jochimstal ein goldschmidt, Caspar Ulich genant; der hat ungeferlich sechszehn stuckh roch goldens erzt, so uns zusteet, bei seinen handen. Und weil wir dann solichs gerne haben wollten, so bevelhen wir euch demnach himit gnedigist, ir wellt unserm münzmaister daselbst im Joachimsthal an unser statt auferlegen, das er soliche stuckh ertz alsbals von ihm dem goldschmidt abfordern und sie euch unser Behaimbischen cammer zu handen wolverwart und also eingemacht, das sie nit schaden nehmen, schickhe, und wann ir die bekhommen habt, sie uns alsdann unverzüglich übersenden." (Fischer, *Kaiser Rudolf II*, 2).

[37] Archiv des Ministeriums des Inneren (heute Österreichisches Staatsarchiv), Kopialbuch 75, fol. 113–4, cit. Katz *Prägemedaille,* 11 note 2.

[38] "nun manglt aber die Schaczung jeczo an dem, dass alhie im Behaim nit dergleichn Goltschmidt und Kunstler so mit solcher Arbeit umbzugehen, und sich darauf verstunden, vorhanden sein, und sollten das die Goldschmidt im Tall, die an denselben Hanndstain etlich Jarlang gearbeitet haben darzue erfordert worden, so ist zuebesorgen, Sie möchten solche in Kunst dermaßen hoch achten und scheczen, dass Euer Kay. Mt. u. Br. dieselb gaar zue theuer ankhomen wurde." (Archiv des Ministeriums des Innern, today Österreichisches Staatsarchiv, Kopialbuch 79, fol. 236–7, and fol. 52r/v).

The *Doppelscheuer*, which might also originate from the Joachimsthal workshops, although it was not created from a single *Erzstufe* (prill) but was assembled from several minerals, must be understood in this context. It is an extremely rare and valuable object, created in a mining area by a goldsmith for aristocratic collectors and their *Kunstkammern*.

These objects were not created by ignorant craftsmen for a disinterested audience, rather they evidence a great sensibility on the part of the maker as well as the recipient. These works of art found their true purpose and destiny in allegorical allusions, in suggested references and complicated reference systems that had to be deciphered gradually by attentive observation. They were catalysts for the pleasure of a learned conservation in which two or more beholders could exchange their sagacious and sharp-witted interpretations.

Surely the learned beholder of the sixteenth century would have first noticed the untreated and rough surface of the coconut. An exterior that in comparison to a contemporaneous 'correctly' accomplished coconut vessel must have been disappointing and might even have been read as a lack of virtuosity in the executive master. But after opening the vessel and discovering the hidden world within, he would have been all the more surprised and astonished by the wit of the goldsmith, who consciously used the natural form as a hint to its 'grown' content. While deciphering the iconography of the inner scenery, the observer may have thought about the forbidden fruit, the expulsion from Paradise, the loss of innocence and thus the loss of the 'uncultivated', i.e. pre-cultural status of humanity.

Not only iconographical programs, such as the reliefs at the foot of the Florentine Campanile (which might go back to a design by Giotto (1266–1337)) or the depictions on the city palace of Ancona, assert this connection between the Fall and the onset of 'artificial' human activity.[39] This connection was often emphasized, especially in texts that were aimed at artists and artisans. An early example is found in the preface to the chapter on painting by Theophilus Presbyter in *De diversis artibus* where he writes:

> We read in the exordium of mundane creation that man, made after the image and likeness of God and animated by the inspiration of the Divine breath, was also, by the excellence of so much dignity, raised above other living creatures; as capable of reason, he merited to participate in the counsel and genius of Divine providence, and, gifted with free-will, he beheld superior to himself but the will of his Maker and the obligation to reverence his decree. Wherefore, miserably deceived by diabolical astuteness, he lost the privilege of immortality through the fault of disobedience, yet so transmitted his power of wisdom and intelligence to his posterity, that whoever would supply care and application might be able to acquire a capability of every art and science, as by a hereditary right. In this manner, human industry, seizing upon this faculty and applying itself in its divers acts to gain and to pleasure, transmitted it, through the development of time, to the predestined epoch of the Christian religion, and it came to pass that a people devoted to God converted to his worship that which Divine ordinance had, to the praise and glory of His name, created. On this

[39] Blume, "Jenseits des Paradieses."

account, the pious devotion of the faithful may not neglect that which the careful prevision of our predecessors transmitted to our age [...][40]

This association of the Fall of mankind and the beginning of cultivation, that is of artisanal human activity, is pointed out even by God, who announces to Adam that from now on he has to feed himself with the work of his own hands (*Genesis* 3, 17–9):

> And unto Adam he said, Because thou hast hearkened unto the voice of thy wife, and hast eaten of the tree, of which I commanded thee, saying, Thou shalt not eat of it: cursed is the ground for thy sake; in sorrow shalt thou eat of it all the days of thy life; Thorns also and thistles shall it bring forth to thee; and thou shalt eat the herb of the field; In the sweat of thy face shalt thou eat bread, till thou return unto the ground; for out of it wast thou taken: for dust thou art, and unto dust shalt thou return.

In the text of Theophilus Presbyter this connection between the human ability of artisanal creativity through his divine derivation (the man as *imago* and *similitudo* of God according to his faculty) is emphasized. But the negative connotations of all human artistic creation as the result of the Fall is turned positively into a *iure hereditario*, into the ability to create art and *ingenium*. For the contemporary beholder the connection of *Kunstkammer* and Fall of Mankind and the correlation of the ability of man to create and cultivate and the loss of the paradisiacal, per definition non-cultivated condition, must have been obvious.[41]

Behind the biblical scene the viewer discovers miners at work, a diviner, who is in search of ore deposits, and two workers with wedges extracting minerals from the surface (Fig. 3). Rotating the coconut vessel, the viewer finds more miners, and the early modern collector might have understood another allusion to the inner coherence of the hard work of ore mining and the Fall of Mankind, two scenes that the goldsmith certainly did not combine by chance in the interior of this coconut vessel. He might possibly also have thought about the scientific advances in mineralogy and metallurgy as part of the cultivating processes that were achieved by humanity since the Fall.

He would certainly have understood, that the divine command not to eat from the forbidden tree of knowledge does not extend to the prohibition of scientific knowledge and progress. The early modern occupation with the possibilities of natural science always included the emphasis that the investigation, the unveiling and understanding of nature might mean a return to the paradisiacal original state, a reversion of the lost Adamic comprehension of the world by just naming it, and that the reading of the book of nature could be the equivalent to reading God's revelation.[42] This becomes evident when Mathesius says that the precise causes

---

[40] Theophilus, *Essay Upon Various Arts* (1847), XLV–XLVII. For the Latin and a German translation, see Theophilus, "Malerei und Glas," 49.

[41] Harrison, *Fall of Man*.

[42] "Doch die Klagen, dass man etwas nur 'mit großer Mühe' bewältigt habe, begegnen einem in erzählenden Quellen nach wie vor so häufig, dass man als innerweltlichen hochmittelalterlichen Hauptantrieb zur Schöpfung neuer Technik die erhoffte Erleichterung körperlicher Arbeit

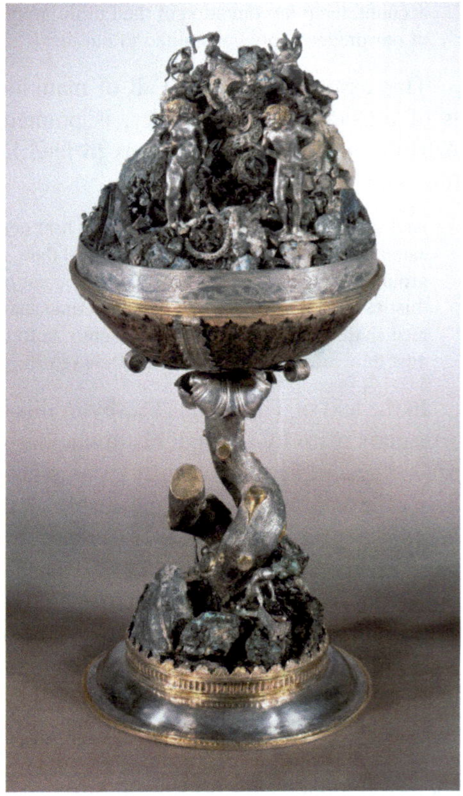

Fig. 3 *Kokosnuss-Doppelscheuer* (*coconut vessel*), open, with the detail of Adam and Eve, sixteenth century St. Joachimsthal (?). Vienna, Kunsthistorisches Inv. No. 885/886

of the origins of the metals is known to God alone, "which we must leave unexplored in present times, until we could see again with new and scrubbed eyes the substantial form of creation that Adam saw before the fall."[43]

Within the discussion of the different conceptions of metallogenesis, the erudite observer would have been able to decipher another layer of meaning, the joke of the coconut. By hiding a huge mountainous landscape within a small coconut the virtuoso goldsmith emphasized the seed-like nature of the fruit. Furthermore, by combining the nut with the mountain he alluded to theories that metals grow from semen in the earth as in a female uterus, thus adding a microcosm-macrocosm analogy to the object. Maybe the scholarly conversation would then have changed to the subject of God as craftsman who, like his earthly colleague, is dependent on good materials and hot fires to make excellent stones and minerals. According to

durchaus erkennen kann. Lehrmäßig ließ sich auf diese Weise ein Sieg über die Sündhaftigkeit Adams erzielen." (Ludwig, "Technik," 36).

[43] "[. . .] die wir wohl noch zur Zeit unerforscht lassen müssen, bis wir mit neuen und gescheurten Augen hinein in die wesentliche Gestalt der Kreaturen wie Adam vor dem Falle wieder sehen werden."

contemporary notions not only exotic nuts and fruits grew better in warmer regions, but also gold and valuable gems were more frequently to be found there. Mathesius states for example, that areas in the vicinity of the earthly paradise (i.e. somewhere in the vicinity of Persia) have particularly excellent mineral and gemstone deposits, because in these regions the earth still reverberates in memory of the voice of God.

After a meticulously detailed examination of the serpent's body the beholder would have noticed an analogy between the artificial braided silver and the naturally occurring forms of a massive silver wire, a curious detail where the artist's hand and the formative power of nature enter into a competitive rivalry. A learned reference to contemporary Italian art theory could have followed, that is, to the creation of Man in the image of God and Man's power to know, imitate and control nature. Federico Zuccari (c.1540–1609) for example addresses man as "almost second God," who received from God the ability of an inner design by which he can understand the world and by which, in imitation of Gold and in emulation of nature can create an infinite number of things, so that he can represent a new paradise on earth:

> He would be almost a second God, he [God] also wants to give him [man] the faculty to form within himself a mental inner design, so that he [man] by means of this could recognize all the creatures & to create inside him new worlds, & inside would have [in form of] a spiritual existence & would enjoy all that, what he enjoys outside [in the outer world] as natural existence & and dominates; & in addition by the means of this design, almost imitating God, & emulating the Nature he [man] could produce infinite artificial things which are similar to the natural ones, & and by means of painting, & sculpture, makes us see new Paradises on earth.[44]

According to Zuccari's art theoretical writings, it is the similarity to God that enables both the scientist and the artist to achieve an inner understanding of nature—the scientist through research to unveil nature's secrets and the artist in artworks that imitate nature's processes. This seems to be the basic assumption that underlies all early modern manipulation of and acting with metals: naturally formed minerals are presented in the *Kunstkammern* next to *artificialia* to exhibit not only the mimicry skills of the artists to imitate the natural forms used as models, but also to imitate the natural processes. The discussion of early modern theories on metallogenesis has shown, that the creative process of the human artist and artisan was compared to the natural genesis of the divine artifex. The same primordial matters are available to the goldsmith and the metalworker, to the alchemist and to 'nature'. All four follow the same working processes, only their ability differentiates the final products. The juxtaposition of form and matter, which is thought of as a dichotomy of active force-shaping power and passive shape-receiving material, is of greatest importance with regard to God and his creation, which stands as a model both for the visual artist as well as the alchemist, who tries to recreate natural processes according to his will.

---

[44] Zuccari, *L'idea* [1607], Book I, ch. VII, 14. See also the facsimile reproduction Zuccari, *Scritti d'Arte* (1961), 162.

It is not only difficult; it is almost impossible to compare the content of a written text with the content of an object. But it seems inadequate to narrow early modern forms and methods of scientific display down to 'codified' knowledge: since the 1920s a long line of historians of science have studied how the input of workmen changed the methods of Renaissance natural philosophy towards a more matter-based natural scientific research and 'experiment'. The concept of the 'superior craftsmen', introduced by Edgar Zilsel in 1942, who started to interact with academics and humanists, is just one of the many important texts dealing with the exchange between the *artes mechanicae* and the *artes liberales* from the fifteenth century onwards, a theory which was recently elaborated and confirmed by Pamela H. Smith in *The Body of the Artisan* in 2004 and by Pamela O. Long in *Artisan/Practitioners* in 2011.[45] Again in 2004, Davis Baird's contribution on 'thing knowledge' focussed on a 'philosophy of scientific instruments' and distinguishes within this epistemology of the artificial object three forms in which an instrument can represent knowledge: firstly as model, secondly as product and medium of working knowledge and thirdly as embodiment of 'encapsulated knowledge'. His theoretical system not only applied to scientific instruments, but can also be adapted to other artificially created objects, for instance to the *Handsteine*. In this field, it offers a wide range of interpretative layers in connection to alchemical knowledge:

First, the *Handstein* as 'model': two forms of exemplary representations can be distinguished interpreting the vessel both from a formal and from an iconographical point of view. Formally speaking, the minerals within the coconut can be understood as an allusion to the alchemical theory of the 'seed/semen-origin' of the metals, as proposed by Palissy. The object thus represents the idea of the versed alchemist, respectively of the alchemist in possession of metal-seeds, who would be able to let ores grow. And at the same time it translates this idea into the artificial area, introducing the *artifex*, the goldsmith, as creator of the nut full of minerals in his working progress. In this model, natural, alchemical and artificial creative processes are blended together.

Iconographically speaking the vessel represents a model of the dichotomy of male/female, form/matter: Adam and Eve embody this principle and simultaneously their action indicates the beginning of every human productive action. After the Fall, every human action is *per definitionem* a repetition of natural creative processes. Embedded in the context of the miners and the general subject of minerals and their genesis and extraction, is the idea of God as creator of metals and the human propensity to imitate this processes—whether in alchemical or artisanal ways.[46]

Second, the *Handstein* as product of working knowledge: This object has also a material significance, because it is the product of applied (al)chemical knowledge. Every piece of art emanating from the goldsmith's workshop is the embodiment of chemical lore concerning material properties and agency. It is important to note,

[45] Zilsel, "Roots"; Smith, *Body of the Artisan*; and Long, *Artisan/Practitioners*.

[46] Newman, *Promethean Ambitions*, ch. III, 34–114.

that the procedures applied by an alchemist and by a goldsmith are in large parts the same: both required the know how to purify metals, to isolate them from the mixed mineral forms they are found in and transform them into massive pure metals. Both the alchemist and the goldsmith share the same theoretical concepts, assuming that sulphur and quicksilver are the two original components of all metals. By alloying these metals, by forcing them into forms, both locate their actions within the context of natural metallogenesis and both suppose, that they thereby imitate God and his creative processes. But it is hard to prove that the goldsmith followed the same ideas as the alchemist, because hardly any artisans recorded their approach in written form: the goldsmith's knowledge is working, not (alchemical) codified knowledge; an object, such as the *Handstein* might count as product, evidence and proof, but will always remain speechless and therefore will never entirely serve (in our modern understanding) as an eloquent advocate to support this assumption.

Third, encapsulated knowledge: A work of art such as the *Handstein* is not 'narrative' in the sense that it develops one argument and discusses it in paratactical form. Such an artwork visualizes the variety of knowledge—the theological, the montanistic, the alchemical—in a synoptical way. Rather than developing an argument in a logical manner, it emphasizes the interdependencies of different areas. But this corresponds with the interconnected nature of these fields, as previously outlined: early modern concepts of metallogenesis must be understood as a mixture of religious connotations, their dependency on antique montanistic theories as well as on alchemical lore adopted through medieval Islamic mediation. Beyond that, a work of art communicates in different ways than a text. The object transforms the know how of the artisan as it is performed in the working process and blends it with the lore of the scholar into embodied, visual knowledge. This knowledge, encapsulated in the work of art can then be exhibited in the surroundings of the ruler, in his *Kunstkammer* collection as well as in processions and courtly festivities.[47] It is—apart from instruments to measure natural phenomena, which Baird puts in the focus of his analysis —not 'useful' in the sense that it embodied knowledge which can be used to gain and widen existent knowledge. But neither is it as useless as the modern beholder of such objects might reckon. The object had its place in the early modern *Wissenskultur*, which operates with analogies. It brought together the practitioner, whose actions included alchemical lore, with the theorist, who tried to balance alchemical theories, antique mineral lore and the newly emerging empirical knowledge from the mining areas, with the sovereign. The latter was interested in alchemical and montanistic research for political as well as economic reasons and used *Handsteine* and comparable objects to stage metallogenetical themes at his court.

[47] Bäumel, "Darstellung des Bergbaus," 213–5.

Conclusion

In conclusion I readdress the questions asked at the beginning of this article. All four questions are inextricably linked: the analysis of the coconut vessel reveals a vast number of formal and iconographical references to the metallogenetic theories negotiated in contemporary discourses, which were embedded in the larger context of art-theoretical ideas on the relationship between man and nature as well as between form and matter. This is not surprising, because goldsmiths, alchemist and *Montantwissenschaftler* (mining professionals) moved in the same 'spaces'—locally and discursively. The example of St. Joachimsthal illustrates the importance of the concept of 'trading zones' for the transfer of knowledge between different professional categories, recently brought into focus by Pamela O Long: in this mining centre the persons processing metals drew their knowledge from the same theoretical writings and treat their matter with the same practices: goldsmiths, alchemist and *Prüfer* (assayer) all were employed in the processes of melting, purifying, and re-forming ores.[48] The examples cited from the treatise as well as from the sermon text of Mathesius showed that there was a common *Vorstellungshorizont*, culturally anchored in the Christian faith, that facilitated exchange and discourse among the different professional groups.

With the help of the coconut vessel and comparable objects, by the transformation of the ore into a *Handstein* and the translocation from its natural environment in the context of the court, these metallurgical discourses materialized and could be staged in the present of the ruler, who was—due to political and economic reasons—also interested in montanistic research.

The *Doppelscheuer*—an object between art and science—thus appears to be a link between the mine and the workshop of the goldsmith, between scholarly discourse and the ordered world of the princely *Kunstkammer*.

Bibliography

1990. *Meisterwerke bergbaulicher Kunst vom 13. bis 19. Jahrhundert*, eds. Rainer Slotta and Christoph Bartels. Bochum: Deutsches Bergbau-Museum Bochum.

1997. *Bei diesem Schein kehrt Segen ein: Gold, Silber und Kupfer aus dem Slowakischen Erzgebirge*, eds. Rainer Slotta and Jürgen Labuda. Bochum: Deutsches Bergbau-Museum Bochum.

Adams, Frank Dawson. 1934. The Origin and Nature of Ore Deposits. An Historical Study. *Bulletin of the Geological Society of America* 45: 375–424.

Amico, Leonard N. 1996. *Bernard Palissy: In Search of Earthly Paradise*. Paris: Flammarion.

Agricola, Georg. 1546. *De ortu et causis subterraneorum*. Basel: Hieronymus Froben & Nikolaus Episcopius.

[48] Long, *Artisan/Practitioners*, ch. IV, 94–126, esp. 107–12.

Agricola, Georg. 1956. Schriften zur Geologie und Mineralogie I. *Georgius Agricola. Ausgewählte Werke*, vol. 3, ed. Georg Fraustadt and Hans Prescher. Berlin: Deutscher Verlag der Wissenschaften.

Albertus, Magnus. 1967. *Book of Minerals*, trans. Dorothy Wyckoff. Oxford: Clarendon Press.

Aristoteles. 1955. Meteorologie. *Die Lehrschriften*, vol. 7, ed. Paul Gohlke. Paderborn: Schöningh.

Bäumel, Jutta. 1990. Die Darstellung des Bergbaus im höfischen Fest des 16. und 17. Jahrhunderts. In *Der silberne Boden. Kunst und Bergbau in Sachsen*, ed. Manfred Bachmann, Harald Marx, and Eberhard Wächtler, 213–215. Leipzig: Edition Leipzig.

Beretta, Marco. 1993. *The Enlightenment of Matter: The Definition of Chemistry from Agricola to Lavoisier*. Canton: Science History Publications.

Biringuccio, Vannoccio. 1540. *De la pirotechnia*. Venice: Per Venturino Roffinello.

Biringuccio, Vannoccio. [1540] 1942. *The Pirotechnia of Vannoccio Biriniguccio: The Classic Sixteenth-Century Treatise on Metals and Metallurgy*, eds. and trans. Cyril Stanley Smith and Marta Teach Gnudi. New York: Dover Publications.

Blume, Dieter. 2006. Jenseits des Paradieses. Skulptur und politische Theorie im mittelalterlichen Ancona. In *Das Modell in der bildenden Kunst des Mittelalters und der Neuzeit. Festschrift für Herbert Beck*, ed. Städelschen Museums-Verein, 25–41. Petersberg: Imhof.

Blumenberg, Hans. 1979. *Lesbarkeit der Welt*. Frankfurt/Main: Suhrkamp.

Cawl, Ulrich Rülein von. [1521] 1955. *Ulrich Rülein von Calw und sein Bergbüchlein*, ed. Wilhelm Pieper. Freiberger Forschungshefte. Kultur und Technik 7. Berlin: Akademie Verlag.

Distelberger, Rudolf. 1985. Gold und Silber, Edelsteine und Elfenbein. In *Renaissance in Böhmen. Geschichte, Wissenschaft, Architektur, Plastik, Malerei, Kunsthandwerk*, ed. Ferdinand Seibt, 255–287. Munich: Prestel.

Dufek, Vladimir. 2004. *Johannes Mathesius (1604–1565). Sein Wirken als Bergprediger in Sankt Joachimsthal und sein Beitrag zur montanistischen Literatur*. Prague, National Technical Museum Prague, 7 Nov 2001. http://www-user.tu-chemnitz.de/~fna/13dufek.pdf. Accessed 13 Nov 2013.

Dülmen, Richard van. 2004. Das Buch der Natur. Die Alchemie. In *Macht des Wissens. Die Entstehung der modernen Wissensgesellschaft*, eds. Richard van Dülmen and Sina Rauschenbach, 131–150. Köln: Böhlau.

Dupré, Sven, and Michael Korey. 2009. Inside the Kunstkammer. The Circulation of Optical Knowledge and Instruments at the Dresden Court. *Studies in History and Philosophy of Science* 40: 405–420.

Ercker, Lazarus. [1580] 1960. *Beschreibung der allervornehmsten mineralischen Erze und Bergwerksarten vom Jahre 1580*, ed. and trans. Paul Reinhard Beierlein. Freiberger Forschungshefte. Kultur und Technik 34. Berlin: Akademie Verlag.

Fischer, Walther. 1969. Bergmännisches in der Sarepta des Johannes Mathesius part 1. *Der Anschnitt* 21: 3–9.

Fischer, Walther. 1970. Bergmännisches in der Sarepta des Johannes Mathesius part 2 & 3. *Der Anschnitt* 22: 21–28.

Fischer, Walther. 1971. Kaiser Rudolf II. Mineraliensammler und Mäzen der Edelsteinbearbeitung. *Der Aufschluss* 22: 1–36.

Hanschmann, Alexander Bruno. 1903. *Bernard Palissy der Künstler, Naturforscher und Schriftsteller als Vater der induktiven Wissenschaftsmethode des Bacon von Verulam. Ein Beitrag zur Geschichte der Naturwissenschaft der Philosophie*. Leipzig: Diederich'sche Verlagsbuchhandlung.

Harrison, Peter. 2007. *The Fall of Man and the Foundation of Science*. Cambridge: Cambridge University Press.

Haug, Henrike. 2012. Wunderbarliche Gewechse. Bergbau und Goldschmiedekunst im 16. Jahrhundert. *Kritische Berichte* 40: 49–63.

Huber, Peter. 1995. Die schönsten Stuffe. Handsteine aus fünf Jahrhunderten. *Extra Lapis* 8: 58–67.

Katz, Viktor. 1932. *Die erzgebirgische Prägemedaille des 16. Jahrhunderts*. Prague: Schulz.

Kayser, Petra. 2006. The Intellectual and the Artisan. Wenzel Jamnitzer and Bernard Palissy Uncover the Secrets of Nature. *Australian und New Zealand Journal of Art* 7: 45–61.

Kemp, Martin. 1999. Palissy's Philosophical Pots. Ceramics, Grottoes and the 'Matrice' of the Earth. In *Le origini della modernita*, vol. 2, ed. Walter Tega, 69–89. Florence: Olski.

Klier, Andrea. 2004. *Fixierte Natur. Naturabguss und Effigies im 16. Jahrhundert*. Berlin: Reimer.

Kris, Ernst. 1926. Der Stil Rustique. Die Verwendung des Naturabgusses bei Wenzel Jamnitzer und Bernard Palissy. *Jahrbuch der Kunsthistorischen Sammlungen in Wien, NF* 1: 137–208.

Laube, Stefan. 2010. Wissenswelten sinnlicher Frömmigkeit. Theatrale Antriebsmomente in der Naturanschauung von Bernard Palissy und Jacob Böhme. In *Religion und Naturwissenschaften im 16. und 17. Jahrhundert*, ed. Kaspar von Greyerz, 217–236. Güthersloh: Gütersloher Verlagshaus.

Loesche, Georg. 1909. Johannes Mathesius. In *Die Wunderstadt St. Joachimsthal. Eine Monographie*, ed. Karl Knopf, 55–127. Weipert/Leipzig: Deutschböhmische Verlagsanstalt Sonnenwirbel.

Long, Pamela O. 2011. *Artisan/Practitioners and the Rise of the New Sciences 1400–1600*. Corvallis: Oregon State University Press.

Ludwig, Karl-Heinz. 1992. Technik im hohen Mittelalter zwischen 1000 und 1350/1400. *Metalle und Macht 1000 bis 160*, eds. Ludwig, Karl-Heinz und Volker Schmidtchen. Propyläen Technikgeschichte 2: 11–205. Berlin: Propyläen.

Mathesius, Johannes. 1564. *Sarepta oder Bergpostill sampt der Joachimsthalischen kurzen Chronicken*. Nuremberg: J. vom Berg und U. Newbar.

Mathesius, Johannes. 1897. Hochzeitspredigten. *Ausgewählte Werke*, vol. 2, ed. Georg Loesche, 169–190. Prague/Vienna/Leipzig: Tempsky.

Modersohn, Mechthild. 1997. *Natura als Göttin im Mittelalter. Ikonographische Studien zu Darstellungen der personifizierten Natur*. Berlin: Akademie Verlag.

Newman, William R. 2004. *Promethean Ambitions: Alchemy and the Quest to Perfect Nature*. Chicago: University of Chicago Press.

Nobis, Heribert M. 1971. Buch der Natur. In *Historisches Wörterbuch der Philosophie*, vol. 1, ed. Joachim Ritter, 957–959. Darmstadt: WBG.

Nummedal, Tara E. 2007. *Alchemy and Authority in the Holy Roman Empire*. Chicago: University of Chicago Press.

Palissy, Bernard. 1996. Discours admirables. *Oevres Complètes*, vol. 2, ed. Keith Cameron. Mont-de-Marsan: Editions Interuniversitaires SPEC.

Quellmalz, Werner. 1969. Zur Materialfrage bergmännischer Handsteine der Renaissance. Untersuchungen eines Handsteins aus dem Grünen Gewölbe zu Dresden. *Der Anschnitt* 21: 14–18.

Rothacker, Erich. 1979. *'Das Buch der Natur'. Materialien und Grundsätzliches zur Metaphern-geschichte*. Bonn: Bouvier.

Schiedlausky, Günther. 1951. Bergmännische Handsteine. *Der Anschnitt* 3: 12–17.

Schlosser, Julius von. 1908. Die Kunst- und Wunderkammern der Spätrenaissance. Ein Beitrag zur Geschichte des Sammelwesens. Leipzig: Klinkhardt & Biermann.

Schmeisser, Martin. 2010. Erdgeschichte und Paläontologie im 17. Jahrhundert. Bernard Palissy, Agostino Scilla, Nicolaus Steno und Leibnitz. In *Diskurse der Gelehrtenkultur in der Frühen Neuzeit. Ein Handbuch*, ed. Herbert Jaumann, 809–858. Berlin: De Gruyter.

Shell, Hanna Rose. 2004. Casting Life, Recasting Experience. Bernard Palissy's Occupation Between Maker and Nature. *Configurations* 12: 1–40.

Smith, Pamela H. 2004. *The Body of the Artisan. Art and Experience in the Scientific Revolution*. Chicago: University of Chicago Press.

Storczer, Stephan G. 1992. *Die Handsteinsammlung des Kunsthistorischen Museums in Wien*. Vienna: Diplomarbeit Universität Wien.

Strieder, Peter. 1967. Erzstufe. In *Reallexikon zur deutschen Kunstgeschichte* 5: 408–418. Stuttgart: Metzler.

Suhing, Lothar. 1986. 'Philosophisches' in der frühneuzeitlichen Berg- und Hüttenkunde. Metallogenese und Transmutation aus der Sicht montanistischen Erfahrungswissens. In *Die Alchemie in der europäischen Kultur- und Wissenschaftsgeschichte*, ed. Christoph Meinel, 293–314. Wiesbaden: Harrassowitz.

Theophilus Presbyter. 1847. *An Essay Upon Various Arts in Three Books, by Theophilus, Called also Rugerus, Priest and Monk, Forming an Encyclopedia of Christian Art of the Eleventh Century*, ed. and trans. Robert Hendrie. London: John Murray.

Theophilus, Presbyter. 1999. Malerei und Glas. *Theophilus Presbyter und das mittelalterliche Kunsthandwerk*, vol. 1, ed. Erhard Brepohl. Cologne: Böhlau.

Zilsel, Edgar. 1942. The Sociological Roots of Science. *American Journal of Sociology* 47: 544–562.

Zuccari, Federico. 1607. *L'idea de pittori, scultori et architetti*. Turin: Per Agostino Disserolio.

Zuccari, Federico. [1607] 1961. *Scritti d'Arte di Federico Zuccaro*, ed. Detlef Heikamp. Florence: Olski.

The Laboratories of Art and Alchemy at the Uffizi Gallery in Renaissance Florence: Some Material Aspects

Fanny Kieffer

Abstract The story of the *Uffizi* Gallery, emblematic monument to the Florentine Renaissance, is still oddly unknown. One of the forefathers of modern European museums, they were built by Giorgio Vasari to cater for Cosimo I's public offices, and were later partly transformed into a gallery by Francesco I de' Medici (1541–1587). Laboratories of art and alchemy were placed side by side by the Grand Dukes Francesco I and Ferdinando I de' Medici (1587–1609) to facilitate collaboration between artists and scientists. Goldsmiths, jewellers, cabinetmakers, sculptors, painters, and cutters of semi-precious stones exchanged not only equipment, but also theoretical and technical knowledge with the alchemists who worked in the *Uffizi*. The pieces that survive demonstrate that the style of the objects created there was a direct result of this collaboration. Thanks to the combined study of archival documents and unpublished maps, the artists' workshops and the alchemists' *fonderia* (foundry) can now be located inside the building. Moreover, thanks to an unpublished inventory, we can easily visualise the organisation of the *fonderia* laboratories, their furniture and the tools that were used. After a short historical introduction, this paper focuses on the material aspects of this collaboration: the working processes, the exchange of instruments between the laboratories, their location in the building and the purpose and destination of the art objects produced.

Considered by some the ancestor of modern museums, the *Uffizi* were first built by Giorgio Vasari (1511–1574) to cater for Cosimo I's (1519–1574) public offices. In 1586, just before his death, Grand Duke Francesco I de' Medici arranged for artist's workshops to be located at the *Uffizi* and made the second floor a gallery. His

The following are abbreviated in the footnotes: Biblioteca Nazionale Centrale di Firenze (BNCF); Magliabechiano (Magl.); Archivio di Stato di Firenze (ASF); Guardaroba Medicea (GM); insertion (ins.); Depositeria Generale (DG); Mediceo del Principato (MDP).

F. Kieffer (✉)
Department of Art History, Centre d'Études Supérieures de la Renaissance, Tours, France
e-mail: fannykieffer@hotmail.com

S. Dupré (ed.), *Laboratories of Art*, Archimedes 37,
DOI 10.1007/978-3-319-05065-2_5, © Springer International Publishing Switzerland 2014

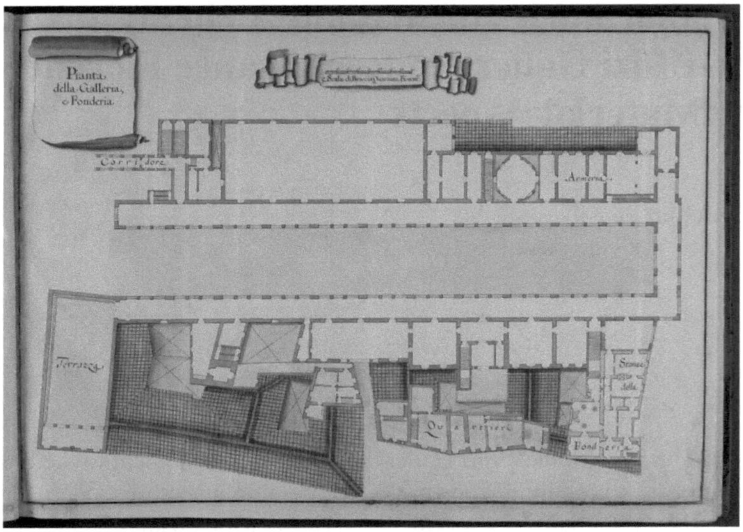

Fig. 1 Map of the second floor of the *Uffizi* with the *Fonderia*, *Piante de' palazzi, giardini, ville et altre fabbriche dell'Altezza Reale del Serenissimo Granducato di Toscana,* Ferdinando Ruggieri. 1742 (manuscript) (Courtesy of Biblioteca Nazionale Centrale Firenze, Florence)

successor, Grand Duke Ferdinando I de' Medici (1549–1609), took up the baton and developed the laboratories and workshops further.

Among the first laboratories to be installed was a *fonderia* (foundry) in the west wing of the building, on the second floor, opposite the garden of the Loggia dei Lanzi (Fig. 1), set up by Francesco I. We may wonder why such bulky installations, including large furnaces and forges, were placed within the *Uffizi* palace. The reasons are quite complex. They involve not only the usefulness of the production processes for the other workshops established in the palace, or reasons related to exclusiveness and secrecy of the production of remedies, but also philosophical reasons related to the encyclopaedic function of the organisation.

In fact, the way the *Uffizi* activities gathered together different types of know how tallies with an encyclopaedic vision of human action on nature. The system of cross-references and correspondences reflect a characteristic aspect of collections unique to this period. Thus, the theatre was close to the musical instruments workshop; the *Stanza delle Matematiche* and the *Stanza delle carte geografiche* contained the maps and scientific instruments built on site, such as the "*macchina universale del mondo*," a large armillary sphere made by Antonio Santucci. The *Tribuna* provided a frame for the most precious objects, natural or artificial: jewellery, precious stones and other *mirabilia* most of which came from the goldsmith's or the *pietre dure* cutters' workshops. In 1588, Ferdinando I set up a big collection of arms and exotic objects in four rooms near the *Tribuna*, placed exactly above the armoury workshops. In same year, Ludovico Buti decorated the vaults of the armoury rooms with scenes of battles and views of the workshops on the first floor (Fig. 2). In the corridors treasured antique and modern statues were

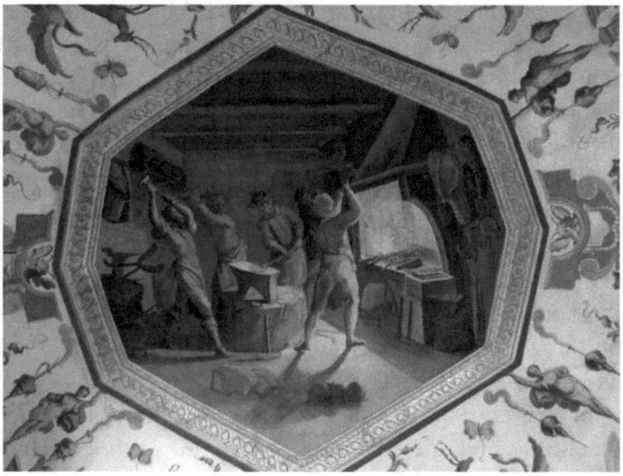

Fig. 2 *Armoury workshops*, fresco, Ludovico Buti. 1588 (Courtesy of the Uffizi, Florence)

displayed under the Giovian series of portraits. In parallel, a restoration workshop for sculptures and a painters' workshop completed the series of know how. Finally, a garden planted with botanic samples on the roof of the Loggia dei Lanzi echoed back to the *fonderia*, situated on the same floor.[1]

According to the archival sources, Francesco I had the original idea of putting together collections and workshops (Ferdinando I's brother and predecessor). He had the necessary philosophical education and cultural knowledge to imagine and execute this complex yet coherent project.[2] From the beginning it was the prince alchemist's brainchild to establish a program gathering *naturalia*, *artificialia* and know how of all kinds under one roof, that would work together like an 'encyclopaedic machine', a kind of monumental clock. The analogy between mechanics and the organisation of the *Uffizi* is not forced: not only were clocks actually built there, but all crafts worked together synchronically.

Artistic objects unique in Europe resulted from this association of technique and aesthetics, and met with great success in other courts. In fact, the uniqueness of the *Uffizi* lies in its administrative and economic organisation: as soon as Ferdinando I became grand duke, he established a new governmental organ, the *Galleria dei lavori*, ruled by a complex hierarchy of ministers, secretaries and intendants.[3] This new institution was completely devoted to the service of the court and the State, it directly served the grand duke's political, diplomatic and economic purposes. Indeed, the *Uffizi* workshops mass-produced specific products for Ferdinando I's

[1] For more on the Uffizi collections in the sixteenth century, see among many others Heikamp, "Geschichte der Uffizien-Tribuna"; Acidini, *Magnificenza alla Corte*; Paolozzi Strozzi & Zikos, *Giambologna*; and Zorzi & Sperenzi, *Teatro e spettacolo*. For a complete bibliography, see Kieffer, *Ferdinando I de Médicis*, 549–82.

[2] See Berti, *Principe dello Studiolo*; and Conticelli, '*Guardaroba di cose*'.

[3] The original document of the *Motu Proprio* from 1588 has disappeared but there is a transcription in Pelli Bencivenni, *Saggio istorico*, vol. II, 119–23.

diplomatic and tributary requirements.[4] In order to best please his allies or to create new alliances, he sent counsellors or spies to other courts (for example, the spy and scholar Filippo Pigafetta, the cardinal Francesco Maria Del Monte, and the musician and art collector Emilio de' Cavalier, who later became the superintendent of the *Galleria dei lavori*) and orientated the production of the laboratories according to their reports.[5] In short, under Ferdinando I, the *Galleria dei lavori* in the *Uffizi* became a real machine at the service of the State.

Economy and profitability were the two main criteria of the administration of the *Galleria dei lavori*. The budget allocated to the court and to the *Uffizi* was very restricted and limited to the indispensable. The workshops were placed in the building in order to rationalize the use of expensive materials and tools. The artists and artisans practiced several disciplines moving freely from one area to another with no hierarchal distinction. A *tinello* (canteen) and accommodation saved time at lunchtime and spared the costs of *vitto e allogio* (food and accommodation) for many employees. The salaries were strictly regulated and the employees could not work for other patrons. This was to try and inhibit knowledge and skills from leaving the court. However, the savings realised on the functioning of the *Galleria dei lavori* were used to buy sumptuous materials: the grand duke did not hesitate to buy the best stones and quantities of precious metals to make the luxurious objects.[6] As a result, the production was very specific and shows the collaboration between all the dependants, artists and scientists. Indeed, the collaboration between painters and *pietre dure* engravers encouraged the creation of new techniques for the painting on stone or for a *commesso* so fine that it looks like painting (Figs. 3 and 4).[7] Sculptors and confectioners joined forces to make sugar statues for banquets; alchemists and painters collaborated on scientific illustration (Fig. 5) and invented new techniques

---

[4] For more on Ferdinando de Medici gift politics, see Butters, "Uses and Abuses of Gifts." Butters concentrates her study on Ferdinando's politics as a cardinal and doesn't focus on art objects, but she gives an idea about the importance, in terms of quantity and meaning, of his gifts. More interesting is Fantoni, "Feticci di prestiggio," and *Corte del Granduca*. See also Mozzarelli, "Onore, utile, principe."

[5] Regarding Ferdinando's artistic politics, I agree with Franco Borsi who says: "[...] anche il rapporto con gli artisti è sostanzialmente cambiato. A quell'intesa preferenziale, a quel binomio che collega come il braccio e la mente Cosimo e Vasari, o Francesco e il primo Buontalenti, e che è una delle forme più singolari e specifiche della storia del Granducato proprio per il coinvolgimento intrinseco della personalità del principe e dell'artista, la convergenza della volontà, lo scambio delle intese, la identificazione psicologica del programma visuale e infine la carica vitale, Ferdinando sostituisce la forma anodina, distaccata, mediatrice ed ambigua del concorso." (Borsi, *Architettura del Principe*, 88). On Filippo Pigafetta, see Pozzi, *Filippo Pigafetta*. On the cardinal Francesco Maria Del Monte, see Wazbinsky, *Francesco Maria Del Monte*. On Emilio de' Cavalieri, see Kirkendale, *Emilio de Cavalieri*.

[6] See Kieffer, *Ferdinando I de Médicis*.

[7] For more on the famous grand-ducal workshops of *pietre dure*, see among others Giusti, *l'arte europea del mosaico,* and *Splendori*; Zobi, *Notizie storiche*; Barocchi & Gaeta Bertela, *Collezionismo Mediceo*; and Acidini *Magnificenza alla Corte*.

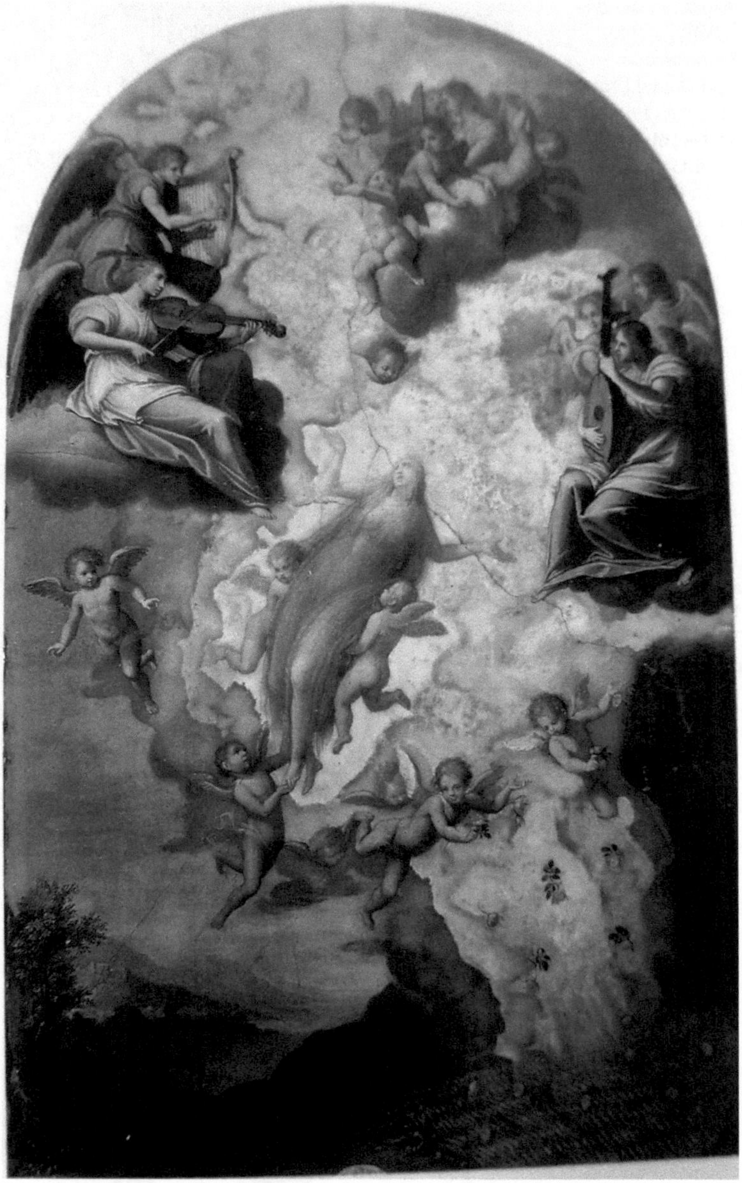

Fig. 3 *Mary Magdalene's Assumption,* alabaster, Valerio Marucelli. End of sixteenth century/ beginning of seventeenth century (Courtesy of Galleria Palatina, inv. Palatina 191, n. 346)

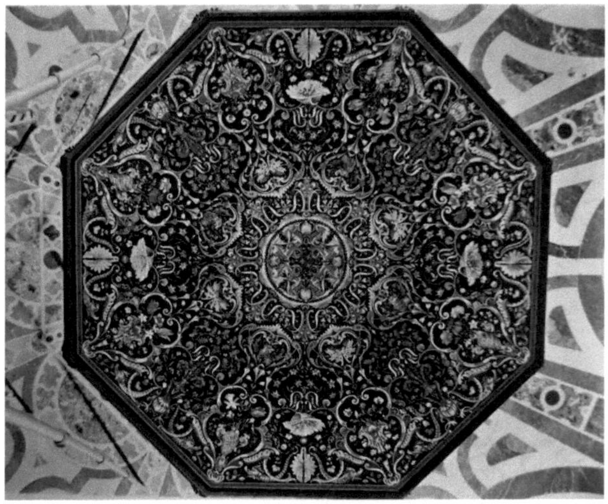

Fig. 4 Table, *pietre dure* on chalcedony, 95 × 84 cm, Jacopo Ligozzi and Daniel Froeschl from Bernardino Poccetti's design. 1597–1604 (Courtesy of Galleria Palatina, inv. Oggetti d'Arte 1911, n. 1512)

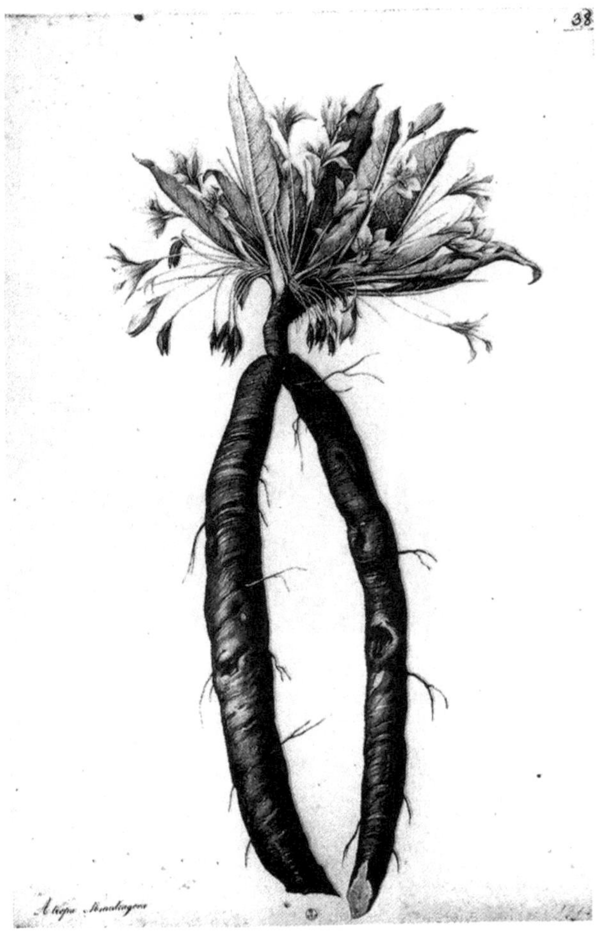

Fig. 5 Mandragora (Atropia Mandragora), Jacopo Ligozzi. c. 1577–1627 (Courtesy of Gabinetto Disegni e Stampe degli Uffizi, 1915 Orn)

for painting; clock-makers worked together with scientific instrument builders to conceive automata or musical instruments for the *Uffizi* theatre.[8]

To our knowledge, no other example of such a governmental system can be found in Europe. Indeed, although most of the European courts were equipped with artistic workshops or laboratories, or even sometimes with both—the courts of Mantua, Prague, Kassel or Munich spring to mind—none of these were subject to such administrative rules and restrictions. Neither did their production serve exactly the same purpose: in most cases, other princes ordered art objects or medicines for their own use and pleasure, for their palace, or for their friends, with no specific political agenda.[9]

The Fonderia in the Uffizi: Origins and Material Organisation

Introduction to the Fonderia

The *fonderia* was made up of a series of rooms on the second floor of the *Uffizi*. In this period a *fonderia* was a laboratory for the arts using fire: metal work and goldsmith activities, alchemy, chemistry, and pharmacy (in this period remedies were made by distillation and extraction of essential oils). There was no precise distinction between the disciplines, for example, alchemy, chemistry and pharmacy are often simply known as 'distillation'. Thanks to the inventories conserved in the archives, we can easily visualise the organisation of the laboratories, their equipment, furnishings and tools.

The first room housed the dispensary, where medicines were weighed and packed before shipment. It was also here that raw material arrived to be transformed in the *fonderia*. The dispensary was fitted with cabinets, a counter with two scales, spoons and funnels, a bench and storage cases. This room was the only one accessible to people who did not belong to the *fonderia*.[10] The three following rooms included a terrace, a 'tower' (a blast furnace) and other furnaces, and housed a complete set of copper distillation tools: eight bains-marie, stoves, plates with covers, basins, bell-jars, mortars and many other small instruments.[11] These were

[8] On scientific illustration in late sixteenth-century Florence, see Ligozzi, *I ritratti*; Tongiorgi Tomasi & Tosi, *"Flora e Pomona"*; Bassani Pacht et al., *Marie de Médicis*; and Garfagnini, *Firenze e la Toscana*, vol. II. Some of those objects conceived by clock-makers and scientific instrument builders can be seen in the exhibition catalogue: Acidini, *Magnificenza alla Corte*.

[9] For more information on the differences and similarities between the *Galleria dei lavori* and the other European court's workshops, see Kieffer, *Ferdinando I de Médicis*.

[10] "Che nessuno pratichi in detta Fonderia eccetto nella prima stanza dove si distribuischono li medicamenti." (ASF, GM 403, ins. 2, fol. 120).

[11] The "terraces" mentioned in the inventory don't exist anymore, but we can deduce from the archival sources they were little open rooms made of wood.

followed by a huge distillation laboratory including an *ordigno* (machine) to distil *acquavite* composed of 60 glass balls, a bain-marie, pallets full of glass pots of all kind and a pierced bench to place the *separatoie* (separators). The following room was full of medicine cabinets and the final room was a smithy including another terrace, equipped with all the tools traditional to this activity (anvil, pincers, sledgehammers etc.).

The Origins of the Fonderia: The Casino di San Marco

At the same time, another *fonderia* in Florence witnessed similar activities. This was the *fonderia* in the *Casino di San Marco*, created by Francesco I long before the one in the *Uffizi*. The visitors and chroniclers of the time gave famous accounts of the experiments Francesco I lead there: for example, he found a way to melt rock crystal and to imitate Chinese porcelain; he made false precious stones, fireworks, and explosives. He experimented with new remedies on dying people, and he practiced alchemy. He did not turn his back on the arts of painting, sculpture, illumination or goldsmithing "and around those things he spends quite all the time in a place called *Casino*, where he has many rooms with masters who do different works and there he keeps his stills."[12]

According to the traditional historiography, the young prince Francesco untiringly frequented Cosimo I's *fonderie*, set up in the *Palazzo Vecchio*, and so recognised in himself the passion that was to prompt him his life long.[13] In fact, as early as 1570, he ordered the architect Bernardo Buontalenti to draw up the plans for the *Casino di San Marco*: he intended to transform it into a palace worthy of a prince and to display there his patronage in arts and sciences. In 1574, when Cosimo I died, Francesco took possession of the *Casino* and of its gardens and set up the first court workshops. Besides the artistic workshops, the palace included a *fonderia* well known for its alchemical research, and its unusual layout. Indeed the suite of adjoining rooms forms a closed circle, a labyrinth that follows a strict structural logic and hermetic aesthetic similar to that of the *Studiolo* in the *Palazzo Vecchio*.[14]

When Francesco I died, in 1587, the usufruct of the *Casino* went to his son, Don Antonio (1576–1621). Ferdinando I, succeeding his brother on the grand-ducal throne, accepted the donation.[15] However, Don Antonio, who was still a child, continued his education at the *Pitti* palace with the other children of the court and

[12] This is the famous ambassador Gussoni's account when he visits the Casino in 1576. Cited in Berti, *Principe dello Studiolo*, 94–5.

[13] See among others Pieraccini, *Stirpe de'Medici*; Berti, *Principe dello Studiolo*; and Barocchi & Gaeta Bertela, *Collezionismo Mediceo*. On the Palazzo Vecchio's *fonderie*, see Perifano, *Alchimie à la Cour*.

[14] About Francesco I's Studiolo in Palazzo Vecchio, see Conticelli, '*Guardaroba di cose*'.

[15] The donation act is reproduced in Covoni, *Don Antonio de' Medici*, 16–8.

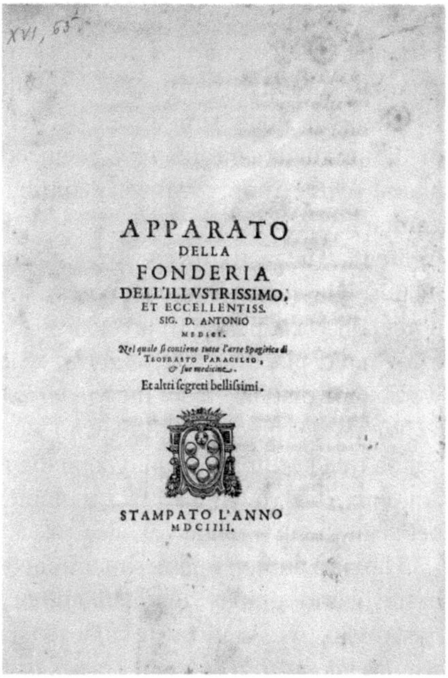
APPARATO
DELLA
FONDERIA
DELL'ILLVSTRISSIMO,
ET ECCELLENTISS.
SIG. D. ANTONIO
MEDICI.
Nel quale si contiene tutta l'arte Spagirica di
TEOFRASTO PARACELSO,
& sue medicine.
Et altri segreti bellissimi.
STAMPATO L'ANNO
MDCIIII.

Fig. 6 Frontispiece, *Apparato della Fonderia dell'Illustrissimo et Eccellentissimo Signor Don Antonio. Nel quale si contiene tutta l'Arte spagirica di Teofrasto Paracelso*. 1604 (Courtesy of Biblioteca Nazionale Centrale Firenze, Magliabechiano XVI, 63, vols. I–IV)

settled at the *Casino di San Marco* only in 1597, at his majority.[16] But the activities in the *fonderia* did not cease in the years between Francesco I's death and Don Antonio's moving, in fact, quite the reverse.

A recipe book preserved at the Biblioteca Nazionale Centrale in Florence, entitled *Apparato della Fonderia dell'Illustrissimo et Eccellentissimo Signor Don Antonio. Nel quale si contiene tutta l'Arte spagirica di Teofrasto Paracelso*, the frontispiece of which was dated 1604, is an eloquent testimony of these activities (Fig. 6).[17] The traditional bibliographic note of this book has always considered the date on the frontispiece and attributed the writing to Don Antonio's entourage.[18] However, a thorough reading of all the recipes reveals a brief note explaining that the writing of the book was undertaken in 1588, during the first year of Ferdinando I's reign.[19] We can deduce that the new grand duke headed two *fonderie*—the one in the *Casino di San Marco* and the other in the *Uffizi*.

For the period we are interested in, two types of sources enable a reconstruction of the activities of the two *fonderie*. For the *Casino di San Marco* the *Apparato*

---

[16] See Covoni, *Buontalenti ai tempi medicei*.

[17] *Apparato,* BNCF, Magl. XVI, 63, I–IV.

[18] Pelli Bencivenni, *Saggio istorico*, vol II, 30; and Galluzzi, "Motivi paracelsiani."

[19] *Apparato,* BNCF, Magl XVI, 63, I, fol. 124.

previously mentioned lists all the projects and recipes worked on in its laboratories. For the *Uffizi*, comparison and cross-referencing of the *Casino* reports with the rare accounts directly from the *fonderia* allow an estimation of its production.

There are many similarities between the two institutions even if the destination of the products and the very essence of their purpose are basically opposed. The layout of the rooms was very similar and both were located near to a botanical garden (even if the one in the Loggia dei Lanzi was very small and its planting very limited). The activities were also the same: distillery of medicinal plants and metals, glass art, pharmacopoeia, smelting and forging metals. Only the art of porcelain was apparently exclusive to the *Casino*. In the *Casino* goldsmithing had a favoured place in the *fonderia*, whereas in the *Uffizi* the goldsmith workshops were some distance from the *fonderia* (although artists forged some items in them).

We do not precisely know who worked in the *Casino*. Nor do the sources reveal which role Ferdinando I played in the research carried out in the *Casino* before Don Antonio's take over in 1597. But numerous clues suggest a permanent collaboration between the two *fonderie*.

The rare author's names mentioned by the *Apparato* are unfortunately unknown in the historiography: the main author "*Giovanni alchemista*" (Giovanni the alchemist) who, speaking in the first person, recorded his own experiences as well as recipes taken from contemporaries or from ancient sources; someone called "G.B." (Giovambattista perhaps) present in 1588; a "famous Lord" Alessandro Cervino; and a "Signor Marcantonio." "Giovanni the alchemist" notes in his book the origin of the recipes. Some of them were Francesco I's doctors, such as Baccio Baldini (already a court doctor under Cosimo I) in charge of the Library in San Lorenzo, or Michelangiolo Angeli da Barga.[20] He also quotes the book of the *Secret of Secret* by the pseudo Aristotle and the Bible.[21] But the name that appears most often is Theophraste Paracelsus, the Swiss doctor mentioned on the frontispiece of the work.

The *Apparato* compiles thousands of recipes and secrets in no obvious order, with no introduction or table of contents. It shows very diverse interests: the explored subjects range from transmutation of metals to chiromancy and from astrology to ballistics. Despite the formal proclamation on the frontispiece of the *Apparato*, Paracelsus was not directly present in the four books: neither are there

[20] For example: "Untione cordiale hauta da Messer Baccio Baldini. Capitolo 397." (*Apparato*, BNCF, Magl. XVI, 63, I, fol. 313). In the roles of 1588 (ASF, DG 389, fol. 10), Baccio Baldini is mentioned as one of the "Medici e spetiali" and also "Messer Baccio Baldini per la cura della libreria di San Lorenzo, scudi 16"; Michelangelo Angeli da Barga is cited among others in *Apparato*, BNCF, Magl. XVI, 63, I, fol. 439: "Unguanto da occhi da Maestro Michelangiolo Angeli da Barga. Capitolo 795."

[21] "Creatione della nuova luna e sole per virtu del zolfo estratto dalla pietra minerale. Capitolo 214." (*Apparato*, BNCF, Magl. XVI, 63, I, fol. 178). The *Secretum secretorum*, also known as *Letter to Alexander*, is a medieval Pseudo-Aristotelian treatise, translated from the Arabic text dated from the tenth century, the Kitâb sirr al-'asrâr. "Giovanni the alchemist" also cites a recipe from the Gospel of Luke recommending to use wine, oil and prayers: "Del modo di medicare con vino, olio e orationi. Capitolo 299." (*Apparato*, BNCF, Magl. XVI, 63, I, fol. 439, and fol. 247).

signs of methodical thought following Paracelsian principles; nor any evidence, even implicitly, of a sense that the Paracelsian concept was completely overturning Galenical medicine. The Swiss scientist's name was only used to suggest the benefits of the chemical arts and as an authority to legitimise the secrets it was used to endorse.[22] Besides, unlike the *Uffizi*, an inventory of the *Casino* from 1621 allows us to reconstruct an extensive chemical library in which around 200 titles are indexed.[23]

Similarities and Differences Between the Fonderie

The general surveyor and "*Maestro di fonderia*" of the *Uffizi* was, till 1587, Michele Geber, of Flemish origin. He seems to have already been in the service of Cosimo I, because Benedetto Varchi mentions him in a manuscript as an author of alchemical recipes.[24] He improved the *fonderia* with instruments taken from the *Casino*. From 1587 to 1589, Geber disappeared from the documents and made way for Niccolò Sisti who lead the operations (his salary increased from 5 *scudi* to 8 *scudi* per month).[25] Sisti had worked before in the *Casino* as a glassblower and a distiller.[26] His background in the *Casino* (as early as 1571, under Francesco I) certainly influenced the production and the methodology of the *fonderia*. In the meanwhile, the *fonderia* of the *Uffizi* was still being fitted out, items were continually being transferred there from the *Casino*, it was growing bit by bit.[27] This indicates that the organisation of the two *fonderie* was basically the same: the same equipment was used, the same persons worked in both institutions. The significant distinguishing feature occurs after the objects produced in the two institutions left the premises.

As soon as Ferdinando I came to power, the *fonderia* of the *Uffizi* played a very important part—if not the most important—in the rationalisation and optimisation process of the production of the court workshops for economic and diplomatic purposes. There was a drop in research and experimentation, because the focus was on production—in some cases almost at an industrial level. This explains the lack of a medical library and also the non-production of treatises or recipe books—apart from dosage notebooks to accompany remedies.

---

[22] About Paracelsianism in Tuscany, see Galluzzi, "Motivi paracelsiani."

[23] "Inventario di tutto quello che si è ritrovato in diverse stanze nel Palazzo detto il Casino dell'Illustrissimo et Eccellentissimo Signor Don Antonio Medici alla sua morte seguita il 2 di maggio 1621. Cominciato il di 3 di detto mese sotto la custodia di diversi ministri che non l'havevono per inventario." (ASF, GM 399).

[24] *Apparato*, BNCF, Magl. XVI, CXXVI, *Alchimia di Benedetto Varchi*, III, fol. 528.

[25] See ASF, MDP 616, ins. 20, fol. 377, and ASF, DG 389, fol. 11.

[26] "Niccolò Sisti luchese che fa con il fiato a lume di lucierna [...] stilatore in Fonderia al Casino da San Marcho." (ASF, GM 183, ins. 18, fol. 47).

[27] ASF, GM 149, fol. 13. See also ASF, GM 183, ins. 3, fol. 79; ins. 5, fol. 50; and ins. 7, fol. 8.

The archives of the laboratory state the current regulations: nobody was allowed to enter the *fonderia*, except in the first room where medicines were distributed; both the work and the orders were to be kept secret; it was strictly forbidden for employees to prepare distillates or medicines outside the *fonderia*; even inside the *fonderia*, no medicine could be prepared or used without advice from a doctor or authority from a grand duke; no medicine could be distributed without a written order from the surveyor of the *fonderia*; all orders were to be registered by the surveyor in a copybook.[28]

The Activities in the Fonderia: Between Art and Science

Glassmaking

The art of glassmaking was well represented among the activities in the *fonderia* at the *Uffizi*. Ferdinando I was the one who decided to transfer glassmakers and their equipment from the *Casino* to the *Uffizi*. During Francesco I's reign the *Casino* housed a very important artistic glass workshop where the prince worked on his own experiments (for example smelting rock crystal).[29] This laboratory was lead by Sisti, who also lead a glassmaking laboratory in Pisa, where he often had to go. In his early career he used the technique *a lume di lucerna*, but once he transfered to the *Uffizi*, this activity became secondary.[30]

At the *Uffizi*, glass production may well have been located in the smithy workshop. It was probably limited to the glass technique *a lume di lucerna*. The oven glass manufacturing method came from existing production centres such as the ones in Pratolino, or Pisa, which became the most important in Tuscany. On the other hand, during Ferdinando I's reign, documents already show activities by someone called Niccolò di Vincenzo Landi di Lucca, after Sisi, the principal glassmaker in service to the Medici's till 1620. The sources show that Landi, a specialist in *a lume di lucerne,* went on duty in the *fonderia* of the *Uffizi* in 1591, where he made little animals for the decoration of glass manufactured in Pisa.[31] In 1601, Antonio Neri also mentioned him as the leader of the new glasswork in the *Casino di San Marco*.[32] Later, in 1618, when Grand Duke Cosimo II (1590–1621) set up a large workshop with many ovens in the Boboli gardens near the Pitti palace,

[28] ASF, GM 403, ins. 2, fol. 120.

[29] On the story of the Medicean glassmaking, see Heikamp, "Mediceische Glaskunst."

[30] The technique *a lume di lucerna* allows to make or decorate little objects heating locally the glass elements thanks to a lantern flame or a candle.

[31] ASF, GM 112, *passim*; GM 217, fol. 23; and GM 195, ins. 1, fol. 102.

[32] On the priest Antonio Neri and his treatise on the art of glassmaking, see Abbri's introduction to Neri, *L'arte vetraria* (2001), 5–23.

he entrusted its management to him, which shows that he was certainly able to supervise large-scale production.

The *fonderia* produced mainly small ornamental glass items (Fig. 7). The items in use for distillery and for the other workshops came either from Pisa, or from the *Casino di San Marco*.[33] Mainly made of bronze or metal, the models of these little items were created by the court goldsmiths or by the smelters: there were little masks, buttons to be filled with perfumes, ornamental glasses.[34] The glasses, stored in a hall of the Galleria "*la stanza dei cristalli*," were distributed according to the grand duke's wishes and other workshops requirements (most of them were given as presents or used to decorate other items coming from other workshops).[35] Sisti, in charge of the *fonderia*, was responsible for the seamless transitions between the different stages of the process, which was not always an easy task in the case of a delicate material such as crystal.[36]

The Forge

In the *Uffizi*, the forge and smelting works were far less visible and have less prestige. Only small items are worked on in the forge: little masks for the ornamentation of furniture, buttons to be filled with perfumes, moulds for crystal, little animals to be worked out in sugar.[37] The forge was mainly used to make objects for the other workshops: locks and keys for furniture, metallic receptacles (cups, buckets, basins etc.), and tools. The leading smith, a Frenchman, Guillaume Lemaître, (Guglielmo di Matre or Lemetre), settled as soon as 1587 in the *fonderia*

[33] ASF, GM 183, ins. 6, fol. 35–36.

[34] ASF, GM 124, fol. 97, 118'; GM 183, ins. 4, fol. 97. For the stylistic aspects of those decorative objects, see Kieffer, "Savant dessinateur"; and Heikamp, "Mediceische Glaskunst."

[35] ASF, GM 183, ins. 18, fol. 47.

[36] "A Maestro Niccolo Sisti a Pisa scrisse il Cavaliere Vinta detto di [6 maggio 1592]. Viene scritto a Sua Altezza Nostro Signore da Siviglia, che tutti quei vetri sono comparsi rotti, et l'Altezza Sua si duole, che spente et poi non ha honore, et tanto piu si maraviglia, che siano rotti questi, perche dovevano andare per mare talche giudica che tutto il difetto nasce dall'essere male incassati, et dovendosene hora mandare di nuovo in Siviglia degl'altri, come dovete sapere, Sua Altezza ricorda che si accommodino con esquesita diligenza, et gli invierete al Proveditore di Livorno, che saprà qualche n'ha da fare, et non essendo questa per altro effetto, mi vi offero. Da Firenze. Al Proveditore di Livorno [Bernardo di Benedetto Uguccioni] scrisse Cavaliere Vinta detto di [6 maggio 1592]. Certi vetri che si dovettono mandare, non è molto in Siviglia, sono arrivati tutti rotti, et perche a Vostra Signoria ne sarà inviata una cassetta da Maestro Niccolo Sisti da Pisa, che ha medesimamente da andare in Siviglia. La Sigoria Vostra ha da inviare a Giovan Antonio o Granduca voglio, con ordine che la mandi al Signor Augusto Titio in Siviglia, con ricordargli che la mandi per mare, che si vorrebbe pure, che una volta arrivassino salvi, et sono tutto al piacere di Vostra Signoria. Da Firenze." (ASF, MDP 280, fol. 128).

[37] ASF, GM 124, fol. 98, 118.

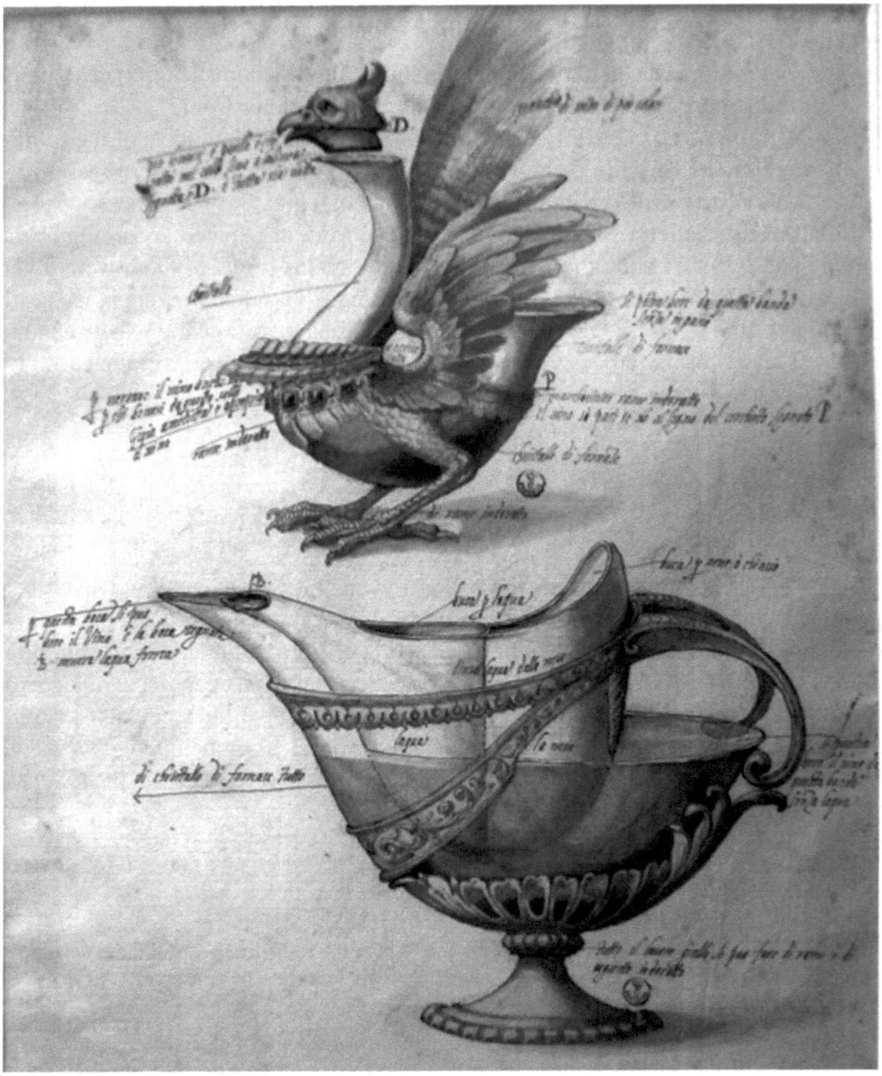

Fig. 7 Glass models, pen, aquarelle, Jacopo Ligozzi. c.1577–1627 (Courtesy of Gabinetto Disegni e Stampe degli Uffizi, inv. 97186)

where the grand duke built him a workshop and a bedroom.[38] The sources document his work and the formation of his workshop until 1601. Taking into account the high quantity of his production, it can easily be imagined that he was essential for the other workshops.

[38] ASF, GM 119, fol. 95.

The Confectionary

The confectionary was allocated to Coriolano Osio, from Verona, who worked in the *Uffizi* from 1587 to 1602. Its function was to prepare jams, marmalades, fruits jellies and mainly sugar figurines, real edible sculptures. The models of these short-lived art works came from the best court artists, such as Jacques Bylivelt or Giambologna and his apprentice Pietro Tacca, who worked out numerous models in bronze and wax for the little sugar animals. These very valuable statues in sugar were reserved for the most prestigious events, such as the wedding of Maria de Medici and Henri IV in 1600.[39]

The supplied models were wax, bronze, plaster, alabaster and even wood (these are produced by the turners). They mostly took the form of little animals, cups, fruit, grotesque niches or lilies. A Venetian bombardier of the Fortezza da Basso was even told to work out models of ships and galleys in wood.[40] Osio turned all these items into coloured sugar in his workshop (which is difficult to situate in the building). These works were often made even more precious with the addition of gilt by the gold beater of the *Uffizi*. Besides sugar, Osio mastered fruit paste work, especially quince paste which was very valued at the court. He was ordered to represent a crowned Florence carrying flowers in its arms with a lion at its feet, from a low relief (probably in wax, perhaps even in stone) manufactured by the sculptors and stonecutters of the adjoining workshops.[41]

Sugar was also a good basis for making medicines: for example, many recipes from the *Apparato della Fonderia di Don Antonio* recommended using syrups, fruit pastes, marmalades or candies to conserve the properties of the active ingredients or amplify their effects. Osio was also in charge of making sugar medicines containing plants and distillations from the alchemical laboratory.[42]

Alchemy

The most important and best-documented activity was by far alchemy (or chymistry) with its various practical applications. The production had several forms, according to the needs of the court: perfumes and cosmetics, remedies of all kinds, poisons and antidotes. In contrast to the production of the *Casino* which, as we saw, was well known due to the writing of treatises and recipes books, the *fonderia*, as far as we know, did not produce any written documentation. So it is only possible to assess the situation through expense forms, orders and letters of instruction to Sisti, the *maestro della Fonderia*, and, above all, through the dosage booklets given with the remedies.

[39] On the confectionary laboratory, see Kieffer, "Confiserie des Offices."

[40] ASF, GM 124, fol. 173; and GM 183, ins. 21, fol. 29.

[41] ASF, GM 124, fol. 209'; and GM 183, ins. 2, fol. 21.

[42] On sugar and medicine, see Kieffer, "Confiserie des Offices."

The dosage booklets are a distinguishing feature of the *fonderia* of the *Uffizi*. They demonstrate that *Uffizi* production was designed to be distributed either as gifts, or as sales products. While the *Casino* concentrated on experimentation and methodological or cultural—even mystical—thought, in the *Uffizi* the recipes have very little variation and the packaging as well as the dosage booklets are mass-produced. For the more famous consignees (such as Cardinal Gioiosa) these dosage booklets are illustrated and sometimes even gilded.

To meet occasional surges in demand, Sisti had to insure a steady supply of raw materials. Sometimes he had to request Cosimo Latini, the minister of the Gallery, to obtain rare or very expensive substances, such as amber or musk.[43] However, most of the raw material came from the Vallombrosa abbey. The monks there grew medicinal plants for the *Uffizi* and collaborated directly with the *fonderia*, as was the case for someone called Giovanni di Giuliano da Montereggi, mentioned as a *erbolaio* or *herbolista* (herbalist) with his assistant Marco di Simone. Both were paid on a daily and merchandise basis, and don't appear in the court roles.[44]

The supply bills coming from Vallombrosa also show which plants are used in the *Uffizi* and for which purpose.[45] We realise that the plant ingredients of certain medicines for the grand duke correspond exactly to the recipes recorded in the *Apparato della Fonderia di Don Antonio*, so that our hypothesis of a collaboration between both *fonderie* during Ferdinando I's reign was reinforced. Here, for example, is a supply bill for herbs from the archives:

> The day 25th of October, in Florence. The Lord Cosimo Latini director of the Gallery. His Grace would be pleased to pay [...] to have brought to His Grace's Fonderia juniper berries to make the oil for the Petechiae Water and to have served for two days mashing the berries [...]. Niccolo Sisti.[46]

And here is its parallel, the recipe from the recipe book of the *Casino di San Marco*:

> His Grace's Petechiae Water. Chapter 237.
>
> Take three ounces of carline thistle, of fine sugar, of cedar pulp, one ounce of cedar seeds [...], juniper berries [...], juniper oil, one ounce of fine theriac, blend and weigh each

[43] ASF, GM 236, ins. 2, fol. 141. See also fol. 143.

[44] ASF, GM 185, fol. 489; and GM 194, ins. 4, fol. 220, 256.

[45] ASF, GM 228, ins. 6, fol. 582; and GM 245, ins. 2, fol. 190.

[46] "Adi 25 di ottobre 1596 in Fiorenza. Magnifico Messer Cosimo Latini proveditore di Galleria. Piacccia a Vostra Signoria di far pagare alli appie l'infrascritte somme per havere consegnato in Fonderia di Sua Altezza Serenissima e per suo servitio, bache di ginepro per farne l'olio per l'Acqua da Petecchie et per havere servito dua giornate a pestare dette bache: A Messer Giovanni da Montereggi e a Marcho da vall'Ombrosa suo compagno, lire sedici et sono per numero sei staia di bachi di ginepro, che hanno consegnato in Fonderia, a lire 213.4 lo staio nette, et lo staio pesa libbre 29 et dette bache servono per trarne olio per fare l'Acqua da Petecchia, lire 26. A Messer Giovanni sopra detto per una giornata che detto ha servito a pestare dette bache, lire 18 a Marcho sopra detto per la giornata d'oggi che serve a pestare dette coccoli, lire 18 Niccolo Sisti." (ASF, GM 193, ins. 2, fol. 166).

thing and put it in fine water for fifteen hours and distil in a well closed glass in bain-marie, use a dose of five ounces each time, half an hour before the fever comes and wave three times.[47]

If we start from the assumption that the recipes recorded in the *Apparato* are also produced in the *fonderia* of the *Uffizi*, they interest us directly in reference to other artistic activities of the *Uffizi* workshops. For example, in the *Apparato*, there are recipes for colouring and fabrication of false gems, for arms and gunpowder, others for how to compose enamel or glass work, especially the technique *a lume di lucerna*:

To repair glasses. Chapter 133.

Take 2 ounces of Saturn glass, 4 ounces of venetian glass, half ounce of borax and put them together in a seal on strong fire, on charcoal, at the first reverberation, and make a glass colour hyacinth, pale, mash it and use this powder to repair with the lamp, blowing the flame in the glass. The Saturn glass is made by putting Saturn in a seal on strong fire, that's to say at the first reverberation.[48]

The court medicines were mainly composed of *Uffizi* manufactured medicines from ancient or tried recipes. The remedies, potions, ointments, powders and other oils were nevertheless uniquely presented with as much grace as refinement: they were placed in ornate glass receptacles, themselves put into very elaborate, painted, gilded, and sculptured ebony boxes with beautiful locks (Fig. 8).[49] In fact, a great

[47] "Acqua da Petecchie di Sua Altezza Serenissima. Capitolo 237. Prendi carlina, zuchero fine, polpa di cedro oncie 3, semi di cedro, di cardo santo, zedoaria, dittamo bianco, terra sigillata, bolo fino oncia 1, reobarbaro eletto oncia 1, foglie di ruta, cinamomo eletto, cassia lignea, coccole d'alloro, oncie 6 d'olio balsamo, carpo balsamo, spigonardi, macis, legno aloe, doronoci oncie 3, seme di ruta, seme santo, seme di portulaca oncie 4, seme di ginepro oncie 2, seme d'acetosa, cicerea bianca oncia 0/2 barbe di tormentilla di gentiana d'angelica galanga oncie 7, olio di ginepro, triaca fine libra 1 mescola et soppesta ogni cosa, et infondi in acqua fine per ore 15 e stilla per vetro ben turato a bagno maria la dose oncie 5 per volta, mezz'ora avanti venga la febre et sventa 3 volte." (*Apparato*, BNCF, Magl. XVI, 63, I, fol. 189).

[48] "Per rassodare i vetri. Capitolo 133. Prendi vetro di Saturno, oncie 2, vetro venetiano buono oncie 4, borace oncia 0/2 et fa correre ogni cosa molto bene in un sigillo a fuoco forte, in su carboni, overo al primo reverbero, e si fa un vetro jacintino, pallido, si pesta, et di quella polvere ci serviamo per rappiccare a una lucerna, soffiando la fiamma nel vetro. Il vetro di Saturno si fa ponendo Saturno in un sigillo a fuoco gagliardo, overo a primo reverbero." (*Apparato*, BNCF, Magl. XVI, 63, I, fol. 130).

[49] "1603. Cassette di sorte d'ebano deono dare addi primo di agosto 1603, n. 1 regia tutta d'ebano di drento et fuora da olij a sepultura, con suo coperchio a cassone et sudetto coperchio si apre per canale a 2 gradi per tenere il libro delle ricette e chiave e nella detta cassetta, n. 24 scompartini grandi e piccoli con 2 mastietti di ferro e sua serratura rabeschati alla zimina a tocco d'oro fine et costa la fattura dell'oro 24.10 scudi, nel corpo detta cassetta una cassetta lungha bassa con 17 spartimenti, tutta d'ebano ed affilettata tutto il corpo e coperchio da tutte le bande d'oro di filo et a provisto la Guardaroba andatocene 70 scudi, auto per le mani di Maestro Gilio Leggi, nel fondo di detto corpo di detta cassetta un'altra cassetta senza spartimenti, da tenere fogli, e tutta con 3 serrature fatte con detti mastietti Maestro Guglielmo Franzese, e si giudica per esser il detto provisionato, costino un suo chiave 10 scudi, lungha 8 2/3, largha 8 ½, alta 8 ½, fatta e fabricata per Maestro Tomaso di Fabbiano e Marchione di Marguett suo compagno, maestri per Sua Altezza Serenissima in Galleria, provisionati, et per avere tenuttone i detti conto del tempo, messoci si dicie esserci di manifattura un mese fra tutta dua per uno 24 scudi, e l'ebano per certo in tutta si valuta 8, a tale che questa cassetta costa in tutta a Sua Altezza Serenissima 115.3.10 e l'inventione e disegno di detta molto bella da sudetti maestri tedeschi." (ASF, GM 261, fol. 10).

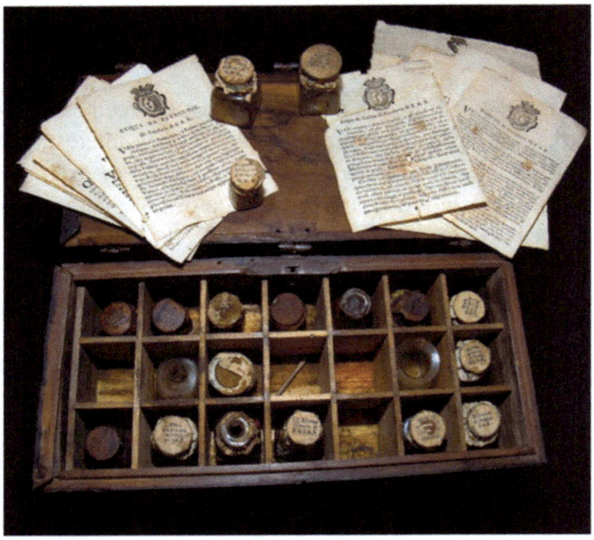

Fig. 8 *Uffizi* workshops, Boxes with medicines and dosage booklets. End of sixteenth century (Courtesy of Museo Storico Nazionale dell'Arte Sanitaria, Rome)

part of the production of the German cabinetmakers' workshop was boxes whose purpose was to contain those remedies. The archives have left us some records of these boxes:

> [. . .] a box covered with red velvet with golden little balls and ribbons made of gold and red silk, lined with red satin, with its silver hinges and inside a lead box and inside there are two little blankets made of taffeta filled with cottonwool, and a necklace of buttons filled with perfume and blackamoors [. . .], there are three ounces of musk and amber, and the remainder was a weight of gold and pearls that was 10.8.0/2 ounces. Given today January 22nd 1593 [1594].[50]

According to this document the boxes were richly decorated with precious buttons and pearl ornamented fabrics, especially when they contained cosmetics and perfumes to be given as a gift to some noble lady. The precious fabrics transformed a remedy box into a piece of jewellery, and this was reinforced by the use of precious gems. The boxes fit one into the other to give another level of surprise. They were divided into compartments containing different medicines, with a compartment for the dosage booklet.

---

[50] "[. . .] una cassetta coperta di velluto rosso con bullette dorate e nastrino d'oro e seta rossa, foderata di raso rosso, con sua gangherature d'argento, entrovi una cassetta di piombo, e in detta vi è dua coltroncini di taffeta rosso imbottiti con bambagina et una collana di profumo di bottoni di mane in fede, e moretti attaccatj che n. 6 bottoni grossi con punte con perlette a fiori, n. 6 bottoni grossi guarniti con rosette d'oro smaltate di bianco, e n. 24 bottoni tondi mezzani guarniti con rosette smaltate, e n. 12 mane in fede con moretti tramezzati con perle tonde, che sono n. 48 perle grosse e n. 22 perle simile, attaccate a moretti che sono delle compere prima, et delle seconde compere ultimamente, che si disse esservi oncie 3 fra musco et ambra, et il resto e il peso dell'oro e delle perle peso tutto oncie 10.8.0/2. Dato adi 22 di gennaio 1593 [1594]." (ASF, GM 185, ins. 4, fol. 327–327').

These boxes, manufactured in great number with their respective dosage booklets—edited by Giorgio Marescotti—were widely distributed all over Europe on every occasion, and had an important artistic and pecuniary value, according to the value of the remedies in them and above all for whom they were intended. In short, they represent a concentration of the know how of the court workshops and their distribution was a good way to promote Tuscan art. Finally, because their efficacy was widely accepted and praised, they were an excellent means for Ferdinando I to create a regular, dependant and indebted clientele.

As a result, the *fonderia* played a double role: it manufactured finished products (the remedies) to be distributed and it produced raw materials for the other workshops of the *Uffizi*. This second part was very important for Ferdinando I's project, because the workshops were arranged so that they could collaborate, one with the other, avoiding the need to bring expensive and indubitably less well adapted materials from outside.

"For knowledge itself was a power whereby he knoweth."[51] The Medici always considered knowledge the key to power in politics. The Grand Duke Cosimo I had already given decisive impulse to all areas of scientific activities within the framework of state centralized structures. For example, he reorganized the *Studio* in Pisa by bringing in famous teachers. Gabriele Falloppio and Realdo Colombo taught anatomy (after Vésale's short stay), Luca Ghini the study of herbs and Giovanni Argentario taught medicine. In 1567, at the expense of the doctors and apothecaries's corporation, Cosimo I edited the new *Ricettario Fiorentino,* a corrected version of the one from 1498.[52] He also established harsh penalties for those who practised medicine or surgery without obtaining the required titles from a doctors' college. Even if Cosimo I's scientific engagement could be placed in the context of an absolutist policy for the sake of prestige, it is also true that he demonstrated a great personal interest in the sciences, which was all the more remarkable as he was not a scholar.[53] In his grand duke's funeral oration, Bernardo Davanzati Bostichi describes a prince with thaumaturgic powers:

> Engines, secrets, oils, distillations, medicine, powerful remedies, because the people from the city but also foreigners and Princes appealed to him with great pleasure, almost as if he were the god Asclepius.[54]

Francesco I and Ferdinando I continued their father's politics. Under Ferdinando I, arts and sciences coexisted in the *Uffizi* to meet the European princess' demands. In all probability the manufacturing and giving of medicines and gifts on a large scale was part of an assimilation strategy by the Florentine

[51] Bacon, "Meditationes Sacrae," vol. VII.

[52] The *Ricettario Fiorentino* is the official handbook for the doctors and apothecaries, containing all the recipes and dosages they have to use. See Lazzi & Gabriele, *Alambicchi di parole*.

[53] See Perifano, *Alchimie à la Cour*.

[54] "Ordigni, segreti, olii, acque, stillamenti, medicine, rimedi potenti, perché a lui con piacer grandissimo quasi allo Dio Esculapio, si ricorreva non pur da quei della città, ma da forestieri eziamdio, e da Principi." (Targioni, *Selve*, vol. VI, fol. 178).

grand dukes towards the thaumaturgy kings of France or England who they envied, while more generally promoting of the arts and sciences of Tuscany. But it would be inappropriate to remain fixed on the idea that the production was only for economic ends. Ferdinando I was without a doubt a pragmatic monarch and concerned with efficiency, but he did not neglect the cultural or artistic aspects of his strategy. For example, in Pisa, he relocated his garden and medicinal *fonderia* in a more suitable place and added a gallery where he stored part of the Florentine collections: natural and extraordinary items, witnesses of special 'operations'.[55] He imitated the great monarchs of his time such as Philippe II of Spain who trusted Francisco Hernandez with a scientific expedition to Mexico. Fernando I financed his own scientific expeditions, among them one by the Flemish botanist Joseph Goodenhuys (most often called Giuseppe Benincasa or Casabona) to Crete. Goodenhuys played a primary role in the study, illustration, introduction, and acclimatization of numerous plants and oriental flowers.[56] When he came back to Tuscany he brought many bulbs and diverse vegetal species, as well as splendid boards showing Cretan plants painted by an anonymous German painter (today held in the Biblioteca Universitaria in Pisa). Ferdinando I also kept close relations with the naturalist Ulisse Aldrovandi (1522–1605) from Bologna: the naturalist and the grand duke exchanged medicinal herbs and commissioned artists to sketch and paint samples to complete their collections of scientific illustrations.

The layout of the *Uffizi* laboratories turned out to be one of the most significant initiatives of Ferdinando I's reign. A result of the three first grand dukes of Tuscany's museographical, cultural and political interests, the *Uffizi* were first exploited as a "machine" of power under Ferdinando I. Cosimo I started to build the palace containing the 13 *magistratures* in order to confirm and demonstrate the Medici's capacity to control and organize the State. Francesco I decided to reserve the upper floor for a dynastic gallery and set up the first workshops. Ferdinando I transferred all the workshops from the *Casino di San Marco* and organized the administration of artistic and scientific activities in order to produce characteristic objects he used as tributes to consolidate his politic alliances.

The cohabitation between arts and sciences still corresponded to the ancient way: in the Renaissance, according to the philosophers and the theologians, arts and sciences both belonged to natural philosophy. Besides, botany and medicine became part of the symbolic network formed around the ideal prince's image: the doctor and philosopher prince acting for his people's health and salvation. This image took on a special aura under Ferdinando I, who dealt in large-scale sacred images with curing powers, like the *Miraculous Virgin* of Santissima Annunziata or the *Madonna* of Loreto, and medicine.

Destined to further dynastic and political propaganda, arts and sciences had to reflect grand-ducal power. Therefore, the *Galleria dei lavori* preserved through secrecy the most progressive techniques, developed a widely recognizable style for

[55] See Galluzzi, "Motivi paracelsiani."

[56] Tongiorgi Tomasi & Tosi, *"Flora e Pomona,"* 18–9.

works of art and employed internationally renowned artists. Ferdinando I organized the production like a huge factory. The objects produced were offered or sold throughout Europe, whether they were works of art or medicines from the *fonderia*. The gifts played a crucial part in diplomatic relations, especially with the Vatican and Spain, promoting the magnificence of Tuscan art.

The great diffusion of Florentine artworks in the European courts also provoked an emulation phenomenon. Rudolf II (1583–1612), Francesco Maria II della Rovere (1549–1631) and Maria de Medici (1575–1642) were tempted to reproduce the *Galleria dei lavori* system in their palaces in Prague, in Pesaro and in Paris. Rudolf II already maintained good relations with Francesco I, punctuated by artistic exchanges. Those exchanges continued under Ferdinando I's reign. Based on the same principles as Francesco I's *Studiolo*, Rudolf's *Kunstkammer* showed many similarities with Ferdinando I's *Uffizi*. Francesco Maria II Della Rovere also drew his inspiration from Ferdinando I's artistic policies. Though he installed artists' workshops and scientific laboratories in his ducal palace in Pesaro, he kept the production for his private use, as did Francesco I at the *Casino di San Marco*. The only transfer that was close to the Florentine model was the one Maria de Medici, Francesco I's daughter, applied to the Grande Galerie of the Louvre. Indeed, she also set out court workshops whose layout and organization was very similar to that of the *Uffizi*, even though the reasoning behind the creation of the set was very different. At the Louvre, the workshops provided the basis for the Académie Royale de Peinture et de Sculpture.[57]

At the *Uffizi*, the workshops activities began to decline on the death of Ferdinando I in 1609. His son, Cosimo II (1590–1621), not without difficulty, tried to maintain the prestige of the *fonderia*. But on his death his successor, Ferdinando II (1610–1670), took immediate measures to decrease the workshop's budget.[58] From this date on, the history of the workshop is unknown. In the Lorraine government reform of 1737, only the *pietre dure* workshop remained—the *Opificio delle Pietre Dure*—and the *fonderia*, which by then consisted only of a pharmacy and a cabinet of naturalist objects.

[57] For the exportation of the Uffizi model in Europe, see Kieffer, *Ferdinando I de Médicis*. See also Bassani Pacht et al., *Marie de Médicis*; Garfagnini, *Firenze e la Toscana*; Fock, "Pietre dure at Court"; Giusti, *l'arte europea del mosaico*; Goldberg, "Artistic Relations"; Marrow, *Maria de Medici*; Montevecchi, "Francesco Maria II"; Neumann, "Florentiner Mosaik"; Schepelern, "Princely Collectors"; and Somers Cocks, *Princely Magnificence*.

[58] "Volendo Sua Altezza rimoderar le spese della Galleria, tanto per quello che si lavora in Firenze come ne l'Arsenale di Pisa, ha resoluto che in futuro non si spenda più di ducati dodicimila l'anno da cominciarsi al primo di febbraio proximo avvenire, però il Cavaliere Vincenzo Giugni soprintendente di essa, consideri bene in quel che sia meglio impiegarli per servizio dell'Altezza Sua e per tutto li 20 di gennaro proximo gnene dia relazione in scrittis, a ciò possa risolvere la sua volontà. Intanto vadia giornalmente licenziando qualcheduno, come segatori e gente simile per non aver in un giorno solo a farlo di gran numéro. E scriva al proveditore de l'Arsenale quanto bisogni e se la prefata Altezza commettessi spesa che ecedesse la sopradetta somma, non si eseguisca se insieme con l'ordine non sarà fatto il mandato del danaro. Lorenzo Usimbardi." (ASF, GM 332, ins. 3, fol. 260, 10 January 1621).

Bibliography

1604. *Apparato della Fonderia dell'Illustrissimo et Eccellentissimo Signor Don Antonio. Nel quale si contiene tutta l'Arte spagirica di Teofrasto Paracelso, & sue medicine. E altri segreti bellissimi*. Florence: Biblioteca Nazionale Centrale di Firenze, Magliabecchiana XVI, 63, vols. I–IV.

Acidini, Cristina. 1997. *Magnificenza alla Corte dei Medici. Arte a Firenze alla fine del Cinquecento*. Milan: Electa.

Bacon, Francis. [1597] 1859–1864. Meditationes Sacrae. Of Heresies. In *The Works of Francis Bacon*, vol. 7, eds. James Spedding, Robert Ellis, and Douglas Heath. London: Longman and Co.

Barocchi, Paola, and Giovanna Gaeta Bertela. 2002. *Collezionismo Mediceo e Storia Artistica*. Florence: Studio Per Edizioni Scelte.

Bassani Pacht, Paola, Thierry Crépin-Leblond, Nicolas Sainte Fare Garnot, and Francesco Solinas. 2003. *Marie de Médicis, un gouvernement par les arts*. Paris: Somogy.

Berti, Luciano. 1967. *Il Principe dello Studiolo. Francesco I dei Medici e la fine del Rinascimento fiorentino*. Florence: Edam.

Borsi, Franco. 1980. *L'Architettura del Principe*. Florence: Giunti Martello.

Butters, Suzanne. 2007. The Uses and Abuses of Gifts in the World of Ferdinando de' Medici (1549–1609). *I Tatti studies. Essays in the Renaissance* 11: 243–354.

Conticelli, Valentina. 2007. *'Guardaroba di cose rare et preziose': Lo Studiolo di Francesco I de' Medici. Arte, storia e significati*. Lugano: Agorà.

Covoni, Pierfilippo. 1892a. *Don Antonio de' Medici al Casino di San Marco*. Florence: Tipografia cooperativa.

Covoni, Pierfilippo. 1892b. *Il Casino di San Marco costruito dal Buontalenti ai tempi Medicei*. Florence: Tipografia cooperativa.

Fantoni, Marcello. 1991. Feticci di prestiggio: Il dono alla corte medicea. In *Rituale, Cerimoniale, Etichetta*, eds. Sergio Bertelli and Giuliano Crifò, 141–163. Milan: Bompiani.

Fantoni, Marcello. 1994. *La corte del Granduca. Forma e simboli del potere mediceo fra Cinque e Seicento*. Rome: Bulzoni.

Fock, Cornelia Willemijn. 1982. Pietre Dure Work at the Court of Prague: Some Relations with Florence. *Leids Kunsthistorisch Jaarboek* 1: 259–269.

Galluzzi, Paolo. 1982. Motivi paracelsiani nella Toscana di Cosimo II e di Don Antonio de' Medici: alchimia, medicina, "chimica" e riforma del sapere. In *Scienze, credenze occulte, livelli di cultura*, ed. Paola Zambelli, 189–215. Florence: Olschki.

Garfagnini, Giancarlo. 1980. *Firenze e la Toscana dei Medici nell'Europa del Cinquecento*, 4 vols. Milano: Electa.

Giusti, Annamaria. 1988. *Splendori di Pietre Dure. L'Arte di Corte nella Firenze dei Granduchi*. Florence: Giunti.

Giusti, Annamaria. 1992. *Pietre dure: l'Arte Europea del mosaico negli arredi e nelle decorazioni dal 1500 al 1800*. Turin: U. Allemandi.

Goldberg, Edward L. 1996. Artistic Relations Between the Medici and the Spanish Courts, 1587–1621. *The Burlington Magazine* 138: 105–114.

Heikamp, Detlef. 1963. Zur Geschichte der Uffizien-Tribuna und der Kunstschränke in Florenz und Deutschland. *Zeitschrift für Kunstgeschichte* 26: 193–268.

Heikamp, Detlef. 1986. Studien zur Mediceische Glaskunst. Archivalien, Entwurfszeichnungen, Glaeser und Scherben. *Mitteilungen des Kunsthistorischen Institutes in Florenz* 30: 1–423.

Kieffer, Fanny. 2007. Jacopo Ligozzi, savant dessinateur de verres. *Journal de la Renaissance* 5: 151–168.

Kieffer, Fanny. 2013. La Confiserie des Offices: Art, sciences et magnificence à la cour des Médicis. *Predella* 33:89–106.

Kieffer, Fanny. 2015. *Ferdinando I de Médicis (1587–1609) et les Offices. Création et fonctionnement de la Galleria dei lavori*. Turnhout: Brepols.

Kirkendale, Warren. 2001. *Emilio de Cavalieri 'Gentiluomo Romano'*. Florence: Olschki.

Lazzi, Giovanna, and Mino Gabriele. 1999. *Alambicchi di parole: Il Ricettario fiorentino e dintorni*. Florence: Polistampa.

Ligozzi, Jacopo. 1993. *I ritratti di piante di Iacopo Ligozzi*, ed. Lucia Tongiorgi Tomasi. Pisa: Pacini.

Marrow, Deborah. 1982. *The Art Patronage of Maria de Medici*. Ann Arbor: UMI Research Press.

Montevecchi, Benedetta. 2001. 'Arti rare' alla corte di Francesco Maria II. In *Pesaro nell'età dei Della Rovere*, vol 2, 323–334. Venice: Marsilio Editori.

Mozzarelli, Cesare. 1980. Onore, utile, principe e stato. In *La corte e il "Cortegiano"*, vol. 2, *Un modello europeo*, ed. Adriano Prosperi, 241–253. Rome: Bulzoni.

Neri, Antonio. [1612] 2001. *L'arte vetraria*, ed. Ferdinando Abbri. Florence: Giunti.

Neumann, Erwin. 1957. Florentiner Mosaik aus Prag. *Jahrbuch der kunsthistorischen Sammlungen in Wien* 53: 157–202.

Paolozzi Strozzi, Beatrice, and Dimitrios Zikos. 2006. *Giambologna, gli dei, gli eroi*. Florence: Giunti.

Pelli Bencivenni, Giuseppe. 1779. *Saggio istorico della Real Galleria di Firenze*, vol. 2. Florence: Per Gaet. Cambiagi stamp. Granducale.

Perifano, Alfredo. 1997. *L'Alchimie à la Cour de Côme Ier de Médicis: savoirs, culture et politique*. Paris: Honoré Champion.

Pieraccini, Gaetano. 1924–1925. *La Stirpe de'Medici di Cafaggiolo: saggio di ricerche sulla trasmissione ereditaria dei caratteri biologici*, vol. 2. Florence: Vallecchi.

Pozzi, Mario. 2004. *Filippo Pigafetta consigliere del principe*. Vicenza: Biblioteca Civica Bertoliana.

Schepelern, H.D. 1985. Natural Philosophers and Princely Collectors: Worm, Paludanus and the Gottorp and Copenhagen Collections. In *The Origins of Museums: The Cabinets of Curiosities in Sixteenth and Seventeenth-Century Europe*, eds. Oliver Impey and Arthur MacGregor, 121–127. Oxford: Clarendon.

Somers Cocks, Anna. 1980. *Princely Magnificence: Court Jewels of the Renaissance, 1500–1630*. London: Debrett's Peerage.

Targioni Tozzetti, Giovanni. 1769–1779. *Selve di notizie, spettanti all'origine de' progressi e miglioramenti delle scienze fisiche in Toscani, messe insieme dal Dottor Giovanni Targioni-Tozzetti, per uso del dottor Ottaviano suo figlio*, 17 vols. Biblioteca Nazionale Centrale di Firenze, Targioni. Tozzetti, Ms. 189.

Tongiorgi Tomasi, Lucia, and Alessandro Tosi. 1990. *"Flora e Pomona": L'orticoltura nei disegni e nelle incisioni dei secoli XVI–XIX*. Florence: Olschki.

Wazbinsky, Zygmunt. 1994. *Il Cardinale Francesco Maria Del Monte, 1549–1626*, 2 vols. Florence: Olschki.

Zobi, Antonio. 1841. *Notizie storiche riguardanti l'Imperiale e Reale Stabilimento dei lavori di commesso in pietre dure di Firenze*. Florence: Tip. di Felice Le Monnier.

Zorzi, Elvira Garbero, and Mario Sperenzi. 2001. *Teatro e spettacolo nella Firenze dei Medic*. Florence: Olschki.

Material and Temporal Powers at the Casino di San Marco (1574–1621)

Marco Beretta

Abstract Built in 1574 by court engineer and architect Bernardo Buontalenti for Francesco I de Medici, the *Casino di San Marco* represents a unique example of a late Renaissance site of alchemical research, art collecting and political court. Francesco I's program to enhance the chemical arts and make it into a body of highly sophisticated knowledge was reflected in the architecture of the Casino which hosted a number of laboratories, several of which survived Francesco's premature death in 1587 and remained active until the beginning of the seventeenth century. It was in this building that the bulk of the first and most successful treatise on glassmaking, Antonio Neri's *L'arte vetraria* (1612), took shape. On the basis of recent archival research, which has provided fresh evidence on the artists employed in the Casino by Francesco and by his son Antonio and on the artifacts which were produced in the laboratories, this contribution briefly explores the history of the Casino and its role in putting chemical arts at the centre of the Medici's patronage. Galileo's arrival in Florence and his telescopic discoveries did not overshadow the extensive presence of chemical arts that, in fact, survived the impact of Galilean science

At the Origin of a Myth

In late November 1780, the Florentine naturalist Giovanni Targioni Tozzetti (1712–1783) published the first volume of a history of seventeenth-century physical science in Tuscany.[1] A monument of erudition, which benefited from Targioni's

The following are abbreviated in the footnotes: Biblioteca Nazionale Centrale di Firenze (BNCF); Magliabechiano (Magl.); insertion (ins.).

[1] Targioni, *Notizie degli aggrandimenti*.

M. Beretta (✉)
Dipartimento di filosofia e comunicazione, Università di Bologna, Bologna, Italy
e-mail: marco.beretta@unibo.it

S. Dupré (ed.), *Laboratories of Art*, Archimedes 37,
DOI 10.1007/978-3-319-05065-2_6,

systematic perusal of the scientific manuscripts preserved in the library of the Grand Duke, the *Notizie* was a deliberate act of historiographic propaganda. Using Galileo (1564–1642) as both a point of departure and of arrival for Tuscan science, Targioni proposed a teleological *tour de force* of 1752 pages to his readers, the most comprehensive historical survey of seventeenth-century Tuscan science ever attempted. Targioni's narrative was filled with documentary evidence and his main argument, that the features of Tuscan science had been dominated by the innovative experimental method set forth by Galileo, gained such authority that his selection of sources remains canonical to this day. The myth of Galileo, already guaranteed by the foundation of the Accademia del Cimento, found its documentary 'evidence' in Targioni's selective history. However, as Targioni himself candidly admitted in the preface to the first volume, he was planning to complement this first version with a more thorough historical account, beginning with the Etruscans, to illustrate the long-term intellectual fertility of Tuscan science and technology. With this bold survey, Targioni intended to explore and emphasize the role played by the Medici in creating the institutional foundation of modern science, in both Florence and Pisa, thus providing Galileo with the ideal cultural background. The bulk of this monumental work, entitled *Selve di notizie, spettanti all'origine de' progressi e miglioramenti delle scienze fisiche in Toscana, messe insieme dal Dottor Giovanni Targioni-Tozzetti, per uso del dottor Ottaviano suo figlio*, reached 17 thick manuscript volumes. A modest abridgment of the work, numbering just over 300 pages, was published in 1852.[2] The principal reason why the manuscript version of Targioni's history was not published is to be found in its contents. Against the wishes of its author, these revealed that Galileo's vision of science had been exceptional rather than the norm and that the interest in the natural sciences shown by the Medici family had been directed mainly to disciplines such as astrology and alchemy, views hardly compatible with Targioni's enlightened values. Although his aspiration to produce an objective narrative led him to include a comprehensive reconstruction of the development of occult sciences during the reigns of Cosimo I (1519–1574) and Francesco I (1541–1587), his disdain for what he dismissed as *fandonie chimiche* (chemical nonsense) undermined his aim of celebrating unreservedly the Medicean dynasty as patrons of 'modern science'.[3] In his account of the life of Francesco I, Targioni remarked that "the dark shadows of alchemy were dissipated [in Tuscany] by the blazing light of the Galilean school," but he was forced to admit that many scholars and enthusiasts for alchemy insisted on collecting manuscripts with alchemical content, and that these went on to form a remarkable body of works, useful to later historians of science.[4] Targioni's

[2] Targioni, *Notizie degli aggrandimenti.*

[3] Targioni, *Notizie sulla storia*, 257.

[4] "Le tenebre dell'alchimia furono dissipate nel nostro paese dalla sfolgorante luce della scuola Galileiana, ma siccome per lo avanti vi erano stati di continuo molti studiosi, ed appassionati per tal arte, non è maraviglia se fra i codici manoscritti di tutte le nostre librerie, si trovano bensì libri di chimica d'autori d'ogni secolo e d'ogni paese, che troppo lunga e noiosa cosa sarebbe il volergli qui registrare. Ve ne sono però molti specialmente sotto il nome di Raimondo Lullo e d'altri

criticism is not surprising and this because of two main reasons: he shared with his contemporaries the view that alchemy and astrology had nothing to do with science; he exalted the fame of Galileo to underline the validity of his wiggish historiography aimed at demonstrating the Florentine origin of modern science.

Targioni's contrasting view of Tuscan science proved to be an influential one. In recent literature devoted to the Medici and their patronage of the natural sciences, the genealogy of scientific disciplines is often presented as having been dominated, since the sixteenth century, by mathematics, perspective and the mechanical arts.[5] This provided Galileo, on his arrival in Florence, with the ideal cultural setting to pursue his work as well as the foundation of the Accademia del Cimento. It is often argued that this picture is evidenced by the Medici's creation of a variety of public and semi-public places, such as the *Uffizi's Stanzino delle Matematiche and Tribuna,* the *Sala delle Carte Geografiche* in the Palazzo Vecchio, and the *Accademia del Cimento* at Palazzo Pitti; here and elsewhere the hierarchy of knowledge they wished to promote was displayed to distinguished visitors.[6] Francesco I's *scrittoio* in the Palazzo Vecchio (the so-called *studiolo*) and his interest in alchemy are portrayed as exceptional cases in the dynasty's history, and the places in which he displayed his scientific interests were not as public as those created by his father and by his successors. We are now in a better position to judge the development of the natural sciences in grand ducal Florence during the late sixteenth century, but we should be aware that the myth of Galileo still casts its hegemonic influence, overshadowing any branch of natural knowledge which seems to deviate from the canons of the new mathematized 'sciences'.

In what follows, I shall try to give a brief glimpse of what is hidden in this shadow. In order to do so we have to look into some older sources, depicting Florentine cultural life just before Galileo's spectacular entrance onto the scene. Moreover, we need to take into consideration that, with the partial exception of medicine, the investigation of natural phenomena was intimately connected with arts and crafts, a prosperous body of activities that, particularly in Florence, had been flourishing since the late thirteenth century. It can be argued that nowhere else do we find such strong ties between the arts and the sciences than in Renaissance Florence. The peculiarity of the Florentine setting had been favored by the extraordinarily rapid elevation of the status of artists, a recognition partly achieved by a growing appreciation for technical innovations and inventions, a development that

solenni maestri d'alchimia, che meriterebbero di esser resi noti; anzi con un poco di pazienza, vi sarebbe da formare un catalogo di aneddoti chimici non spregevole." (Targioni, *Selve*, vol. VIII, 160–1).

[5] See, for instance, Camerota, *I Medici e le scienze.* A partial exception to this dominate trend is the catalogue of the Medici exhibition of 1980, *La corte il mare i mercati.* Here Paola Zambelli has shown the pervasive relevance of alchemy and occult science at the Medici court. Unfortunately Zambelli's section was artificially separated from that of 'science.' On the historiographic distortions originated in the Galileian myth see also Galluzzi, "Motivi paracelsiani."

[6] Heikamp's article of 1970 has set the standard, see Heikamp, "Antica sistemazione."

potentially challenged the prerogatives of traditional academic intellectuals.[7] These achievements were also the result of political influence exerted by the guilds on the government of the Tuscan capital.[8] As early as the fourteenth century the so-called *arti minori* managed to be represented at all levels of the municipal government.[9] It is beyond the scope of this presentation to examine the causes of the expansion of Florentine guilds, but it suffices here to note the remarkable importance achieved by those related to the chemical arts, such as pharmacy, dyeing, glassmaking and goldsmiths.[10] In 1427 there was only one spectacle maker shop in Florence but in 1480 this had increased to seven.[11]

Mapping the Florentine Arts

In 1584 a magnificent map of Florence, made by the grand duke's cosmographer Stefano Bonsignori (?–1589), presented the topography of the city from an isometric perspective (Fig. 1).[12] Tracing the outlines of its physical features, sites and buildings in considerable detail, it showed a city that in little more than a century had been transformed by a new diversified architectural vision. In addition to resplendent churches, palaces and state offices, Florence witnessed the establishment of several sites dedicated to the experimental sciences and to the applied arts connected with them.[13]

In the year 1561 there were 2,182 workshops providing a livelihood for around 10,000 artisans and serving a population of 70,000 inhabitants.[14] The crisis of the Republic doubtlessly contributed to the gradual decline of the guilds which no longer could, and in some cases no longer sought to, restrain the entrepreneurial ingenuity of their more enterprising members.[15] In addition, the ranks of local artists and craftsmen were increased by foreigners who flocked to Florence, confident of finding patrons for their work. Reciprocally humanists, with the experience of a century of close and fruitful collaboration, no longer sought to dictate the

7 Although innovation was not a category appreciated in all Florentine guilds, the innovative role played among others, by Florentine architects, engineers, pharmacists, painters, sculptors, goldsmiths created a favorable context for the cultural and social enhancement of their professions.

8 Staley, *Guilds of Florence*; and Doren, *Arti fiorentine*. For a more recent reconstruction (with updated bibliography) see the collection of essays edited by Franceschi & Fossi, *Arti Fiorentine*.

9 Goldthwaite, "Realtà economico-sociale," and *Building of Renaissance Florence*.

10 Instructive data can be drawn by the recent reconstruction made by Bianchi & Grossi, "Botteghe, economia e spazio urbano."

11 Ilardi, *Renaissance Vision*, 95–115.

12 Buonsignori, *Nova pulcherrimae civitas*.

13 In addition to the *Casino* and the *Galleria degli Uffizi*, several botanical gardens, both private and public are visible on the map.

14 Miniati, "Fabbro sia un buon maestro," 284.

15 The cases of Leonardo and Benvenuto Cellini were not exceptional.

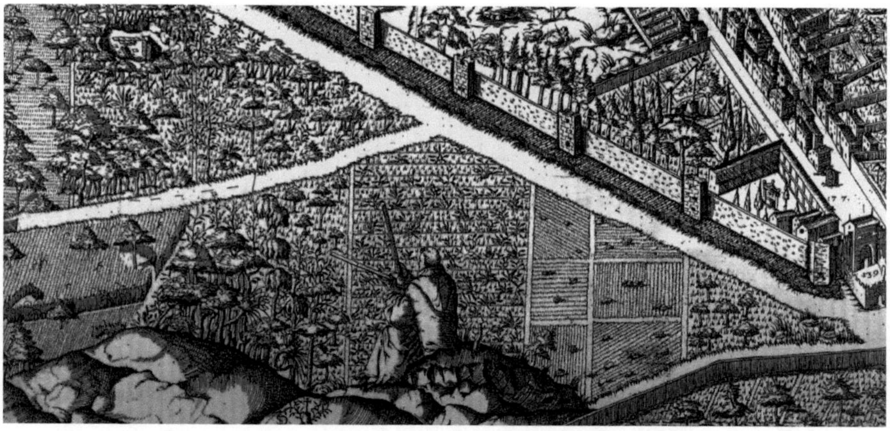

Fig. 1 Self portrait by Stefano Buonsignori in his map of Florence *Nova pulcherrimae civitatis Florentiae topographia accuratissime delineata* (1584). Private collection

canons for artists and artisans, but often interacted with them as equals.[16] Not surprisingly, in 1588 the Grand Duke Ferdinando (1549–1609) defined Florence as a city "founded upon the guilds and trade."[17]

This new configuration of professional relations led to a thorough reassessment of the locations where cultural and economic activities were conducted, articulating in an experimental key the arts, which previously had been regarded as the fruit of individual dexterity rather than the outcome of a deliberate cultural program. In many cases the architectural changes were not apparent, since many kinds of artisans (e.g., the goldsmiths) kept on using the same premises they had occupied for generations. However, even in such examples of apparent continuity the arts were seen as vectors for change. This perception found visual expression in a volume of engravings, based on drawings by Jan van der Straet (1523–1605) (or Giovanni Stradano), that bore the significant title *Nova Reperta*, commissioned by the Florentine academician Luigi Alamanni (1558–1603) around the end of the 1580s (Fig. 2).[18] The book's intertwining of the visual arts, technical inventions and scientific discoveries opened the curtain on a new epoch and Stradano depicted many of the sites associated with the birth of experimental research, beginning with Francesco I de Medici's *studiolo* in the Palazzo Vecchio, whose decoration

[16] The collaboration between Leon Battista Alberti and Filippo Brunelleschi opened a new setting for the social relations between the world of learning and the creative craftsmen. One century later a literary figure such as Benedetto Varchi could take special pride in boasting of his friendships with craftsmen.

[17] "Fiorenza è citta fondata sull'arti e traffichi mercantili," cited in Corazzini, *Diario Fiorentino*, 273.

[18] Baroni Vannucci, *Jan Van Der Straet*, 397–400.

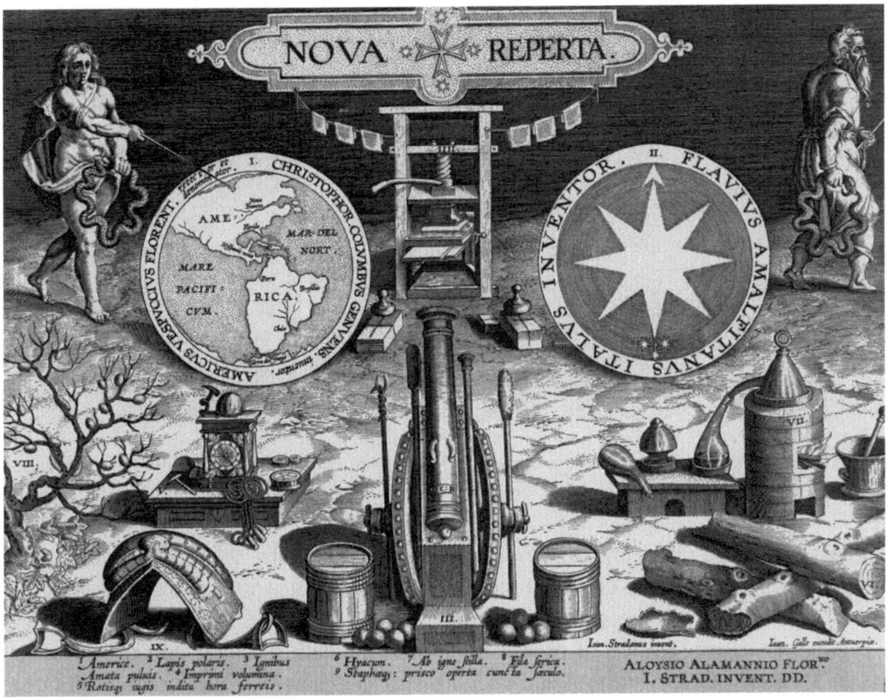

Fig. 2 First engraving and Luigi Alamanni's and Stradano's *Nova Reperta* (1587–1589) (Courtesy Museo Galileo Florence)

reflected the grand duke's alchemical interests.[19] This emblematic space had been preceded by Cosimo I's *fonderia*, a distillery located among a few court workshops and the many display rooms of the Palazzo Vecchio, and would be followed by the network of workshops and laboratories at the *Casino di San Marco*, the *fonderie* of the Uffizi Gallery and Palazzo Pitti.[20] Throughout the city innovative spaces with new purposes were created during this period: natural history museums annexed to botanical gardens filled with rare plants, the ateliers of glassmakers, and the creation of new and more specialized apothecary shops, while traditional institutions such as hospitals and public squares assumed new forms. Even such familiar landmarks as the Loggia dei Lanzi were invested with new significance during the *Cinquecento*; originally the site of public assemblies and ceremonies, the Loggia was now even used on occasion to display natural history exhibits from the Medici collections for the edification of the general public and, housed a botanical garden on the roof.[21] Monographs on some of these sites have been published, most of

[19] Although Stradano's representations of the arts did not explicitly refer to Florence in the captions, several visual details reveal their connection with the Tuscan capital.

[20] On the history of the *fonderie*, see Piccardi, "Fonderia Medicea di Firenze."

[21] Vossilla, "Cosimo I, lo scrittoio del Bachiacca."

them by art historians who have underlined the role of the Medici as patrons of the arts.[22] Important as their work has been in documenting the historic relevance of the new architectural settings, a global picture is still lacking of the nexus between these sites dedicated to the production of scientific and technical knowledge and the economic forces that propelled their realization. A recent book on the history of the *Spezieria al Giglio* illustrates the profound economic and cultural changes introduced by apothecaries who were only apparently immobile.[23] Cosimo I's policy of sponsoring the arts and technology as part of his program of economic expansion forced intellectual elites to take account of a body of knowledge that had been confined to the margin of the Court, and to re-evaluate their room for manoeuver, not only conceptually but also architectonically, in terms of their work spaces.

The Casino di San Marco

In this process of reconfiguring Florentine arts and crafts, alchemy played an exceedingly important role, both as a fashionable scientific discipline with not particularly strong academic ties and as a useful set of experimental practices in which several crafts helped artisans quench their thirst for innovation and realize their socio-cultural ambitions. I shall not explore here the spread of interest in alchemy in Florence at beginning of the sixteenth century, a theme already effectively surveyed in Alfredo Perifano's book on Cosimo I's alchemical passion.[24] I shall instead take into consideration the development of alchemy in the Palazzo Vecchio's *fonderia* and the *Casino*, sites where its relation with the chemical arts and technological innovations became apparent. Indeed, it is my impression that if we want to understand the metamorphosis of the sciences in Florence during this crucial period we should look at disciplines, such as alchemy, which had few academic ties and, at the same time, established close connections with the arts. The connection between alchemical pursuits and the artisanal ateliers of glazers, smiths, goldsmiths and apothecaries favored the introduction of significant innovations in the laboratory or, as it was called at the Medici court, of the *fonderia*.[25]

[22] Berti, *Principe dello Studiolo*; Heikamp, Antica sistemazione, and *Studien*; Butters, *Triumph of Vulcan*; Conticelli, '*Guardaroba di cose rare*', and *Alchimia e le arti.*

[23] Shaw & Welch, *Making and Marketing Medicine.*

[24] Perifano, *Alchimie à la Cour.*

[25] According to Rinuccio Galluzzi, Cosimo's interest in alchemy originated in his efforts to exploit Tuscan mines: "Questo esercizio delle miniere inspirò al Duca il gusto dominante del secolo di formar l'oro con la combinazione di diversi metalli. Siccome le semplici ed evidenti teorie della fisica erano avviluppate nella peripatetica oscurità perciò gli effetti della natura non sapeano investigarli che per vie occulte e straordinarie. Cosimo avendo concepito una singolare inclinazione per questa arte la più vana di tutte eresse nel suo Palazzo una fonderia in cui si compiaceva occuparsi nelle diverse composizioni dei metalli e dei minerali; tutti i segretisti del secolo erano favorevolmente accolti da esso, che godeva di apprendere nuovi metodi per fare

Although important steps in this direction had already been taken during the reign of Cosimo I, it was especially with Francesco I that the reconfiguration of the alchemical laboratory became a central concern. The connection of alchemy with the chemical arts was visually displayed in the series of paintings hanging in the *Studiolo* of Palazzo Vecchio, a small dark room decorated with a complex iconographic narrative. Designed for the then Prince Francesco by Vincenzo Borghini (1515–1580) and Giorgio Vasari (1511–1574), its cabinets seem to have included some of the most valuable treasures produced in the *fonderia*. In addition to all sorts of precious stones, glass and porcelain, they carefully stored the rarest natural specimens and the products of alchemical experiments.[26] The meaning of this room and its intimate connection with Francesco I's interest in alchemy, has been discussed by many, and most recently reconstructed in the comprehensive work by Valentina Conticelli. Thirty-two artists were employed and coordinated by Vasari to accomplish the work. Benvenuto Cellini (1500–1571), who shared with his patron a keen interest in alchemy, was meant to have taken part—but his premature death prevented his involvement. In the *Studiolo* the connection between alchemy and the chemical arts was exalted. The iconographic itinerary conceived by Borghini was situated in a sequence on the upper part of the wall where paintings devoted to thermal baths, the discovery of gun powder, a glass works, a goldsmith's workshop, alchemists and a bronze foundry were located; at the room's two ends were two statues, portraying Vulcan and Apollo. On the lower part of the wall a mythological scene recalled Francesco I's alchemical and artistic interests.[27] I would like to stress the importance of a few elements directly or indirectly related to the chemical arts. What first strikes you in this particular arrangement is the proximity of Stradano's famous painting of Francesco I's alchemical laboratory and Giovanni Maria Butteri's (1540–1606) painting illustrating the glass works (Figs. 3 and 4).

As the design of the *Studiolo* was carefully planned by Borghini together with Francesco, this disposition almost certainly reflects their views on the role of the chemical arts in the reform envisaged by the Grand Duke. Stradano's painting, signed and dated 1570, depicted Francesco's alchemical laboratory in Palazzo Vecchio before it was moved to the *Casino*.[28] The apparatus in the foreground shows a distillation still of the type invented by the Florentine physician Taddeo Alderotti (1215–1295) and described by Vannoccio Biringuccio (1480–1539?).[29]

---

esperienze; la composizione dei veleni non fu l'ultima delle sue ricerche, ed ebbe credito in Italia di fabbricare i più violenti. Siccome gli errori e le vanità qualche volta conducono alla scoperta di cose utili, questa fonderia li rese celebri per l'Europa per i rimedj e medicinali che vi si fabbricarono in progresso." (Galluzzi, *Istoria del granducato di Toscana*, vol. I, 158–9).

[26] Conticelli, '*Guardaroba di cose*', 61–3.

[27] The order and disposition of the artworks in the *Studiolo* has been reassessed in the cited study by Conticelli.

[28] Cosimo I had in fact two *fonderie* in Palazzo Vecchio which according to Vasari were "la fonderia vecchia [...] [e] la fonderia nuova" which may have been situated in the palace's south-east corner. See Butters, *Triumph of Vulcan*, vol. I, 246–7.

[29] Biringuccio, *Pirotechnia*, 128v.

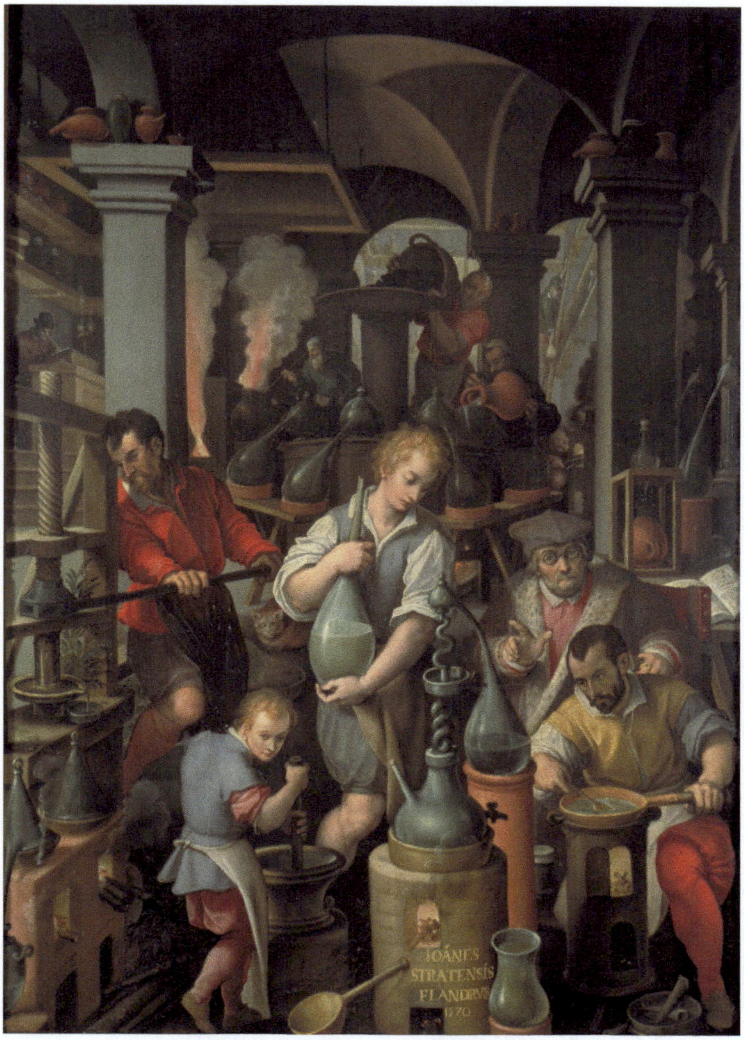

Fig. 3 Francesco I (*right*) de Medici's alchemical laboratory by Stradanus (1570) (Courtesy Museo di Palazzo Vecchio Florence)

The person sitting on the right is Francesco and the alchemist behind him has been identified as Sisto de Bonsisti from Norcia who was employed by Cosimo in the early 1560s thanks to his skill in counterfeiting precious stones and whose son Niccolò eventually would be employed by Francesco to produce crystal glass.[30] A painting close to it, executed by Butteri and dating to 1570–1572, describes a glass works in some detail; in the background we see a glass furnace of the Murano type,

[30] Conticelli, '*Guardaroba di cose*', 335–6.

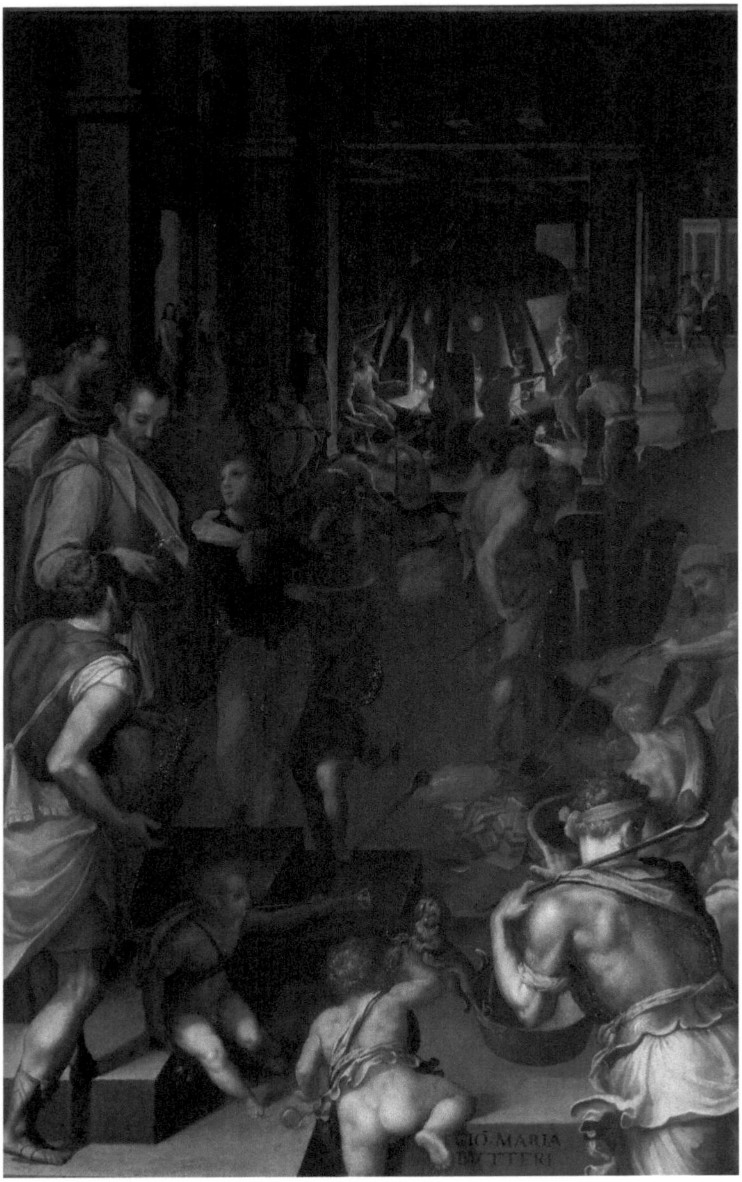

Fig. 4 Francesco I (*left*) visiting his glass work. Giovanni Maria Butteri (1570–1572) (Courtesy Museo di Palazzo Vecchio, Florence)

similar to the one illustrated in Biringuccio's *Pirotechnia* (fol. 43v), where the techniques of shaping glass artifacts both by free blowing and by blowing into a mold are shown. The furnace painted by Butteri suggests that he was illustrating the glass works set up by the Venetian glassmaker Bortolo who arrived in Florence in

1569 when the *Studiolo* was taking shape.[31] This is confirmed by the fact that on the right side of the painting, we see material for the construction of a new furnace and on the left Francesco is examining finished piece of glassware. The size of the glass works is impressive and the entrance shown in the background of the painting with a couple of distinguished Florentine courtiers or citizens just inside the door opens up a glimpse of the *Loggiato* of the *Galleria della Uffizi.*[32] Between Stradano's painting of the alchemical laboratory and Butteri's of the glass works, a painting by Alessandro Fei (1543–1592) depicts a man who seems to be Prince Francesco working on the gem-studded grand ducal crown in the atelier of a goldsmith and jeweler (Fig. 5). The arrangement of the painting suggests that glassmaking and jewelry were the arts most closely connected with alchemy. Their connections inspired Francesco to create a new decorative scenario, dominated by alchemy, by which illustrate the relation between arts and nature.

Francesco was extremely keen, even more than his father, to promote the study of the natural sciences and the arts related to them. His passion for the chemical arts was so strong that in 1560, when he was only 19, he was told off by his brother Giovanni for attending the works of the *fonderia* all day long.[33] It was Francesco who, in September 1569, charged the architect, engineer and inventor Bernardo Buontalenti (1531–1608), who had been one of his teachers since 1550, to oversee the construction of Bortolo's new glass furnace and a new *fonderia.*[34] Francesco I also exploited Buontalenti's versatile skills in the works of the fusion of precious stones and, in particular of rock crystal. Francesco's project to enhance his multi-faceted interest in the alchemical and chemical arts found a new, imposing site with the construction of the *Casino di San Marco* (1567–1574), a sumptuous palace designed by Buontalenti (Fig. 6).[35] Across the street from the Convent of San

---

[31] As pointed out by Heikamp. *Studien*, 63–9.

[32] At the time the ground floor of the *Uffizi* was already completed. I thank Suzy Butters for providing me with this information.

[33] "Guardi di non si profonder troppo nel piacer della Fonderia, che qua vien detto, che ella non esce mai et massimamente il giorno; talché al ritorno nostro speriamo di veder qualche nuova e bella invenzione," cited in Berti, *Principe dello Studiolo*, 28. Berti (p. 51) rightly believes that the *fonderia* was that of Francesco's father in the Palazzo Vecchio.

[34] "ò trovato M. Bortello, et li ò provisto tuti e matoni fra il giardino e a l'artiglieria, tanti che non à più di bisogno: la fornace è a buono porto; et m'a deto che vorrebe che vostra E.I. li facesi paghare qualche danari per potere dare a' maestri che à menati, che ne vorebano mandare a le loro famiglie [. . .]," letter by Buontalenti to Francesco dated 20 Sep, 1569, cited in Heikamp, *Studien*, 344.

[35] On early works on the building see the documents published by Butters, "Pietra eppure non una pietra," 178, 184. The first contemporary description of the building, dating 1591, is that by Bocchi, *Le bellezze della città di Firenze*, 8–9: "Casino, edificato dal Granduca Francesco. Sono in questo palazzo stanze divisate con mirabil arte, in tanto numero, che dentro ogni gran Principe habitar puote adagiato comodamente: ci è una Guardaroba piena di ricchi arnesi, come quadri di preziosi marmi, tavole di diaspri, panni tessuti con singular lavoro, e un letto insino dell'Indie portato a noi di valuta, e di artifizio grandissimo. Il disegno di questo palazzo è di Bernardo Buontalenti, huomo di peregrino ingegno e raro, come si vede nelle finestre, che sono leggiadre, nelle camere, che sono artifiziose, ma quella che è principale in su la strada è bella a maraviglia." On the *Casino*, see Covoni, *Buontalenti ai tempi Medicei*; and Fara, *Bernardo Buontalenti*, 156–65.

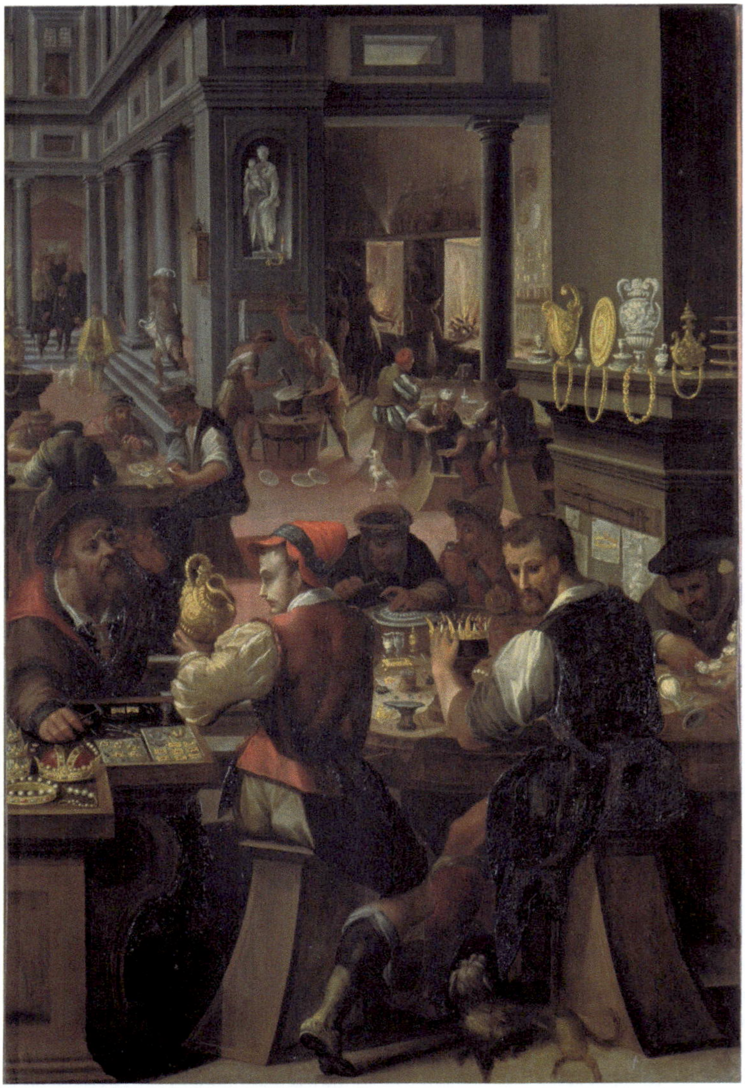

Fig. 5 Francesco I (*right*) working with his father's crown. Alessandro Fei (Courtesy of Museo di Palazzo Vecchio, Florence)

Marco, a complex built in 1442 by the architect Michelozzo Michelozzi (1396–1472) under the patronage of Cosimo il Vecchio (1389–1464), the *Casino* encompassed the *Orti medicei*, a symbolic place commissioned by Lorenzo il Magnifico (1449–1492) to host the excellence of Florence's fine arts (Fig. 7). As early as 1574, when Francesco I inherited the *Orti* from his father he used the existing building as an *officina di esperimenti chimici e fisici*, thus introducing a

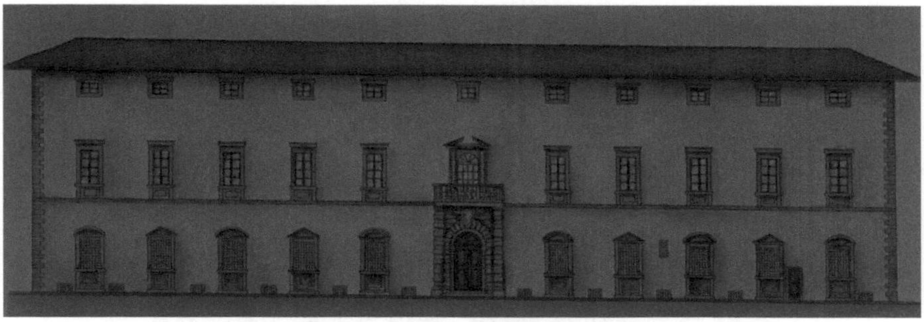

Fig. 6 The *Casino di San Marco* (Courtesy of Museo Galileo, Florence)

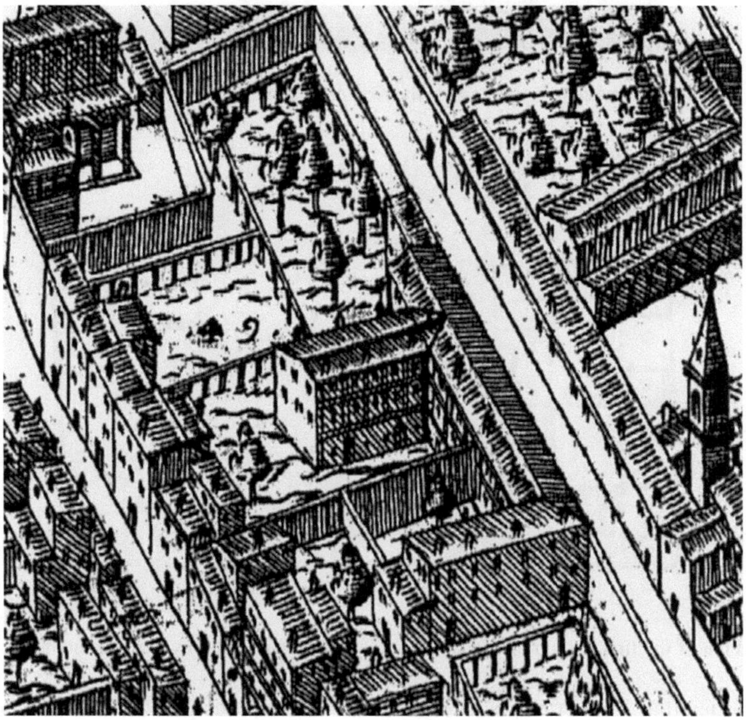

Fig. 7 The *Casino di San Marco* and the *Orti Medicei* in a detail of Buonsignori's *Nova pulcherrimae civitatis Florentiae topographia accuratissime delineata* (1584). Private collection

new hierarchy of the arts.[36] The *Casino* incorporated the old buildings into a new and unusual one, which is particularly evident in the bizarre architectural decoration of the windows (Fig. 8). The building, now a tribunal, has been restructured so

[36] Covoni, *Buontalenti ai tempi Medicei*, 12.

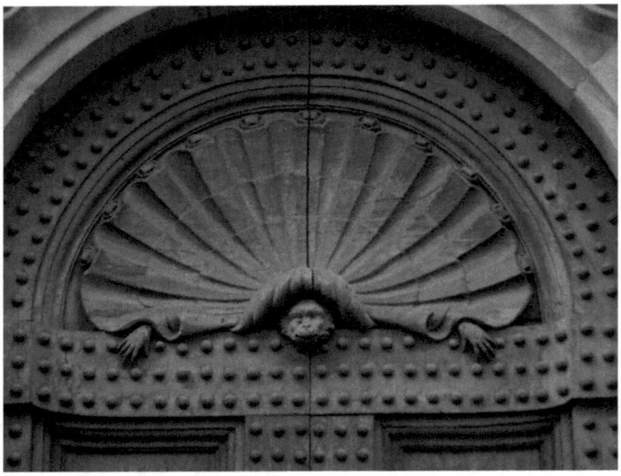

Fig. 8 Detail of Buontalenti's window of the *Casino di San Marco*

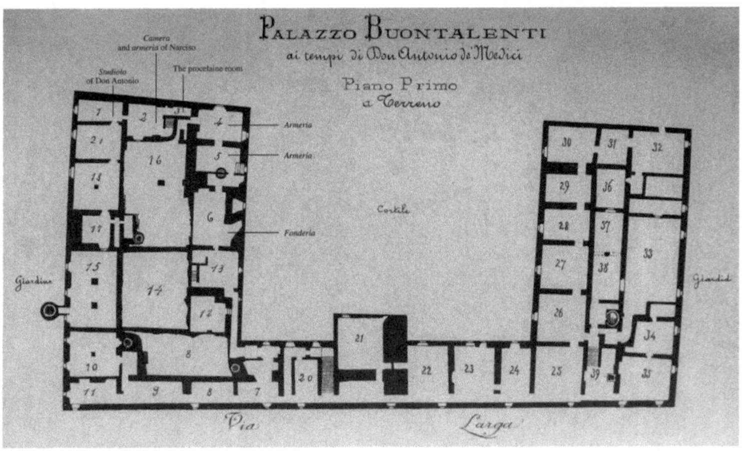

Fig. 9 Plan of the first floor of the *Casino di San Marco* published by Covoni in 1892b

many times that it has progressively lost its original design. As far as I know, no sixteenth-century drawings or engravings of the original disposition of the residential rooms and the workshops survive today. Pierfilippo Covoni found a plan of the first floor of the building post-dating the death of Francesco's son Don Antonio de' Medici (1576–1621). Since we know that Don Antonio restructured the building in 1594 we should be cautious about putting too much weight on this document (Fig. 9). At the beginning of the seventeenth century the *fonderia* occupied a row of first floor rooms; in Don Antonio's day, these were preceded by a library, to which I will return when describing Don Antonio's activities. We certainly know that in Francesco's *Casino* there existed a glass works, an alchemical laboratory and a furnace for producing porcelain: painters, goldsmiths and gem carvers were also active in the *Casino* but it is not clear where exactly their workshops were sited, and how they related one to another. On the basis of the evidence available it seems that Buontalenti and Francesco carefully planned the disposition of the arts, putting the chemical ones at the top of their concerns. In addition to these workshops, Francesco displayed some of his spectacular collections of paintings, sculptures, coins, gems, *naturalia* and *mirabilia* in several of the *Casino*'s rooms and the fountain with Giambologna's sculpture in its garden. If the arrangement of their display is not precisely known, we have several contemporary descriptions which help to provide us with a relatively good picture of what was going on within the walls of the *Casino*. For example, we have several documents reporting on the salaries of the artists and alchemists employed by Francesco at the *Casino*.[37] In addition to Buontalenti, who coordinated the activities of the various laboratories, the salaried artisans of the *Casino* in 1580 included the following 18:

> 'Tanai de' Medici proveditore de' tapezieri scudi 4 [tapestry]
> Maestro Antonio portoghese tappeziere, scudi 10 [tapestry]
> Ieremia Foresti fonditore, scudi 12 [metal worker]
> Niccolò di mastro Sisto, scudi 5 [Medici 'porcelain', glass, alchemy]
> Filippo della Serena mastro de vetri, scudi 12 [glass]
> Giovanni Ambrogio milanese, scudi 20 [silver and rock crystal]
> Stefano milanese, scudi 20 [silver and rock crystal]
> Giuseppe che lavora con li duoi milanesi, scudi 6 [silver and rock crystal]
> Pier Maria detto il Faenzino, scudi 10 [Medici 'porcelain' and majolica]
> Giuseppo da Campo stovigliaio, scudi 7 [Medici 'porcelain' and majolica]
> Giuseppe Marchesi vineziano, scudi 30 [ruby maker and carver]
> Giorgio milanese intagliatore di cammei, scudi 25 [carver of cameos]
> Cristofano figliuolo del suddetto Giorgio, per il primo luglio 1578, scudi 16 [carver of cameos]
> Iacopo Ligozza veronese, scudi 25 [painter and illustrator]
> Messer Giovanni Battista Framberti mantovano, per il primo di agosto 1579, scudi 35 [alchemist]
> Messer Ardicino Castelletti, per il primo di agosto 1579, scudi 35 [alchemist]
> Messer Niccolino Merli, per il primo di agosto 1579, scudi 20 [alchemist?]
> Maestro Buonaventura Rocchigiani, scudi dugento l'anno, cominciati il dì primo di febbraio 1579, scudi 16.4.13.4.[38]

Important as it is, this document is not comprehensive because in other periods Francesco I relied on at least three masters for glassmaking: Bortolo from Murano, Niccolò Sisti from Norcia (supervisor of the *fonderia*) and Buontalenti. Furthermore, he employed the Milanese goldsmiths Giorgio Gaffuri, Ambrogio and Stefano Carono to produce luxury rock crystal vases, the Venetian carvers Salvatore Pocatena, Lorenzo Capogrossi and Giuseppe Marchesi to work on precious stones, the Dutch goldsmith Jacques Bylivelt (1550–1603), several alchemists such as the archbishop Antonio Altoviti (1521–1573), Giovan Battista Framberti,

[37] See, for instance, the correspondence published by Conticelli, "Lo Studiolo di Francesco I."

[38] Stipendiati del Casino San Marco (1580) (Archivio di Stato di Firenze, Mediceo del Principato 616, ins. 20, fol. 377) published in Barocchi & Bertelà, *Collezionismo mediceo*, doc. 178, 163–4. It is interesting to note that after becoming Grand Duke, Ferdinando changed his brother's policy and, with the exception of Giambologna, he fired most of the alchemists and made efforts to uniform the salaries of the artisans employed by the court.

Ardicino Castelletti, Nicolino Merli, Ettore Barbisoni, Sebastiano Manzoni, sculptors such as Giambologna (1529–1608) and Cellini, scientific illustrators such a Jacopo Ligozzi (1547–1627), and many painters who, as we have already seen, had been coordinated by Borghini and Vasari in the realization of the *Studiolo* and who eventually continued to work in the *Casino*.[39] In 1580, one of the grand duke's alchemists at the *Casino* reported that he had 12 large furnaces at work, including one distilling vinegar.[40]

In 1576 the Venetian Ambassador in Florence, Andrea Gussoni (1546–1615) reported, with a sense of wonder, the grand duke's dedication to the chemical arts: among the inventions attributed to the Grand Duke Gussoni listed the production of soft paste porcelain, fusing rock crystal (which was attributed by Vasari to Buontalenti), counterfeiting precious stones (especially emeralds), carving precious stones, preparing several pharmaceutical remedies for distribution to both the Florentine people and foreign rulers, producing fireworks (again developed by Buontalenti) and a new method for the multiplication of saltpeter. In addition to these activities, which absorbed much of his time, Francesco was portrayed by Gussoni as a connoisseur and collector of paintings, sculptures, miniatures, cameos, medals and all sorts of antiquities, of the types displayed in the *Casino*.[41]

Gussoni also reported that Francesco not only employed several skilled artisans at the *Casino*, but that he also personally performed all sorts of chemical experiments, especially those concerned with glassmaking. Ulisse Aldrovandi's (1522–1605) travel diary, dated 1577, also reports on Francesco I's experimental practices. After having studied natural history in Luca Ghini's (1490–1556) botanical garden in Pisa, Aldrovandi became one of the most distinguished late Renaissance Italian antiquarians and naturalists and from the late 1560s on, he enjoyed the patronage and friendship of Francesco I, with whom he shared, among other things, an interest in scientific illustration. During his visit in Florence in the spring of 1577, Aldrovandi described in some details the collection displayed at the *Casino*, among which he noted an emerald vase, made of colored glass and rock crystal, and other precious stones.[42] From Aldrovandi's account we learn that Francesco's delight in rock crystal and precious stones was shared by many other Florentine collectors, such as the apothecaries Gori Bamberini and Stefano Rosselli, and learned men such as Niccolò Gaddi (1537–1591), Anton Maria Salviati (1537–1602), Giorgio Soderini and Francesco Malocchi. Moreover, Aldrovandi gives us a long list of alchemical books for sale in the workshop of the main Florentine

[39] For Francesco's patronage of philosophy literature and, more generally, the academic world see Berti, *Principe dello Studiolo*, 43 ff.

[40] "Tutti i dodeci forni grandi ritrovandosi in lavoro che danno quarantotto vasi et altri sei di rame, che stillano l'ultimo accetto, che riportassi da parte conforme al buon volere di Vostra Altezza," letter by Giovanni Battista Franberti to Francesco I (24/08/1580), published in Conticelli, "Lo Studiolo di Francesco I," 242.

[41] Albèri, *Relazioni degli ambasciatori veneti*, 376–89.

[42] Aldrovandi, "Itinerarium seu rerum in itinere Florentino, Romano et Tyburtino collectarum catalogus," in Tosi, *Ulisse Aldrovandi*, 213.

typographer Giunti, which testifies to their local popularity. In a letter addressed to the Grand Duke dated September 1577, Aldrovandi recalls the *Casino* as a *casa di natura* where miraculous experiments were performed. He mentions Francesco's 'porcelain' vases, which exceeded in beauty and value the Emperor Nero's *vasa muhrrina* mentioned in Pliny's *Natural History*, the fusion of rock crystal which led to the production of most beautiful artifacts, the secret of making steel tools hard enough to temper porphyry and lapis lazuli, as well as many *prestantissimi secreti*.[43] In the *Casino* Francesco also collected books and manuscripts related to secrets. Among them we find numerous recipe books, many from Venice, containing instructions for the production of counterfeit precious stones, crystal glass and all sorts of remedies, including many alchemical.[44] For nearly a decade the Casino was certainly among the most important buildings in Medici Florence.[45] Francesco used his palace both as a place where he could promote technical innovations in the arts and, after he became Grand Duke in 1574, as a focal point for Medici political power, a place where he received ambassadors, aristocrats and intellectuals. In addition to those already mentioned, Cardinale Luigi d'Este (13 June 1581); the Archduke Massimiliano (18 November 1581), the Nuncio of Spain, Monsignor Taverna (16 October 1586) were most impressed by the building and its novel arrangements.

Guests to the *Casino*, whatever the reason for their visit, were introduced to the workshops' most luxurious and innovative products, and to the grand duke's collections. Indeed, the *Casino* was a uniquely hybrid site in which politics, science and the arts were all part of a synergetic strategy, a strategy which struck the visitors with a sense of surprise and wonder. Not even in Rudolph II's Prague was a connection like this so explicitly embodied in one site.[46] While Cosimo I kept politics and his alchemical interests separate, Francesco united them in the *Casino*. Incidentally it is worth noting that because of this unusual combination, Francesco would eventually be portrayed as an ineffective ruler, with an inclination to melancholy.[47]

In a letter to the Duke of Urbino dated 12 July 1586, Simone Fortuna claimed that the *Casino*, a *bellissimo e grandissimo palazzone*, was built by Francesco for his "Principe Antonio", the much favored illegitimate son of the Grand Duke and

[43] Tosi, *Ulisse Aldrovandi*, 239–40, 246.

[44] These include a recipe book dating from 1585, partially drawn from the activities of the *fonderia* of Francesco, where there are recipes for coloring rock crystal to resemble topaz, emerald and sapphire (BNCF, Magl., XV, 142, fol. 155v–158v, cited in Targioni Tozzetti, *Selve*, vol. VIII, fol. 18).

[45] See also the report on the *Casino* made by Del Riccio, *Istoria delle pietre*, 171.

[46] Francesco and Rudolph II were acquainted and during their youth had resided in 1562 at the court of Philip II in Spain. On Philip's interest in alchemy see Bueno, "Mayson pour Distiller Eaues." Philip's interest in this occult science, however, developed after Francesco's stay in Madrid.

[47] Berti, *Principe dello Studiolo*.

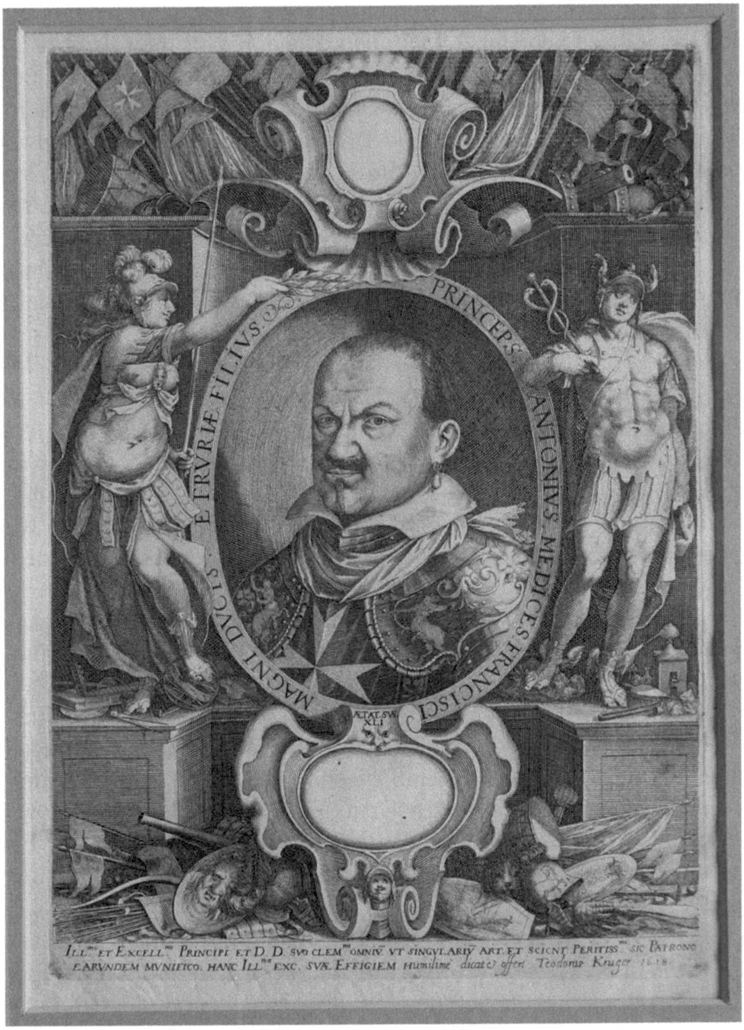

Fig. 10 Engraved portrait of Antonio de' Medici at the age of 41 (1618). Th. Kruger (Courtesy of Museo Galileo. Florence)

Bianca Cappello (1548–1587).[48] At the time the letter was written Antonio was not yet 10 years old, and only 1 year later, in October 1587, Francesco and Bianca would both die. In spring 1594, upon Antonio's foregoing any future claim to the grand ducal succession, the new Grand Duke Ferdinando endowed him with the *Casino* together with several villas, palaces and estates.[49] In order to make the

[48] Barocchi & Bertelà, *Collezionismo mediceo*, doc. 320, 287.

[49] The list of which is reported Covoni, *Don Antonio de' Medici*, 40–1; and Parigino, *Tesoro del Principe*, 137–45.

Casino his residence, Antonio de' Medici, then 18, began to renovate the palace and to refurnish it with a rich collection of artworks and *naturalia*, but it was only in 1597, that he was able to move in (Fig. 10).[50] It is not clear whether any of the chemical workshops continued functioning between the death of Francesco and Ferdinando's endowment. It is well known that Francesco transferred many of the workshops to the *Galleria degli Uffizi*, and that his alchemists were fired by Ferdinando but it may well be that some of their laboratories were left at the *Casino* so that their existing facilities could be exploited. A few of them were still in existence when Don Antonio took over the *Casino* because he immediately put them to work.[51]

Little is known about Antonio de' Medici's cultural background: Giovanni Battista Paggi (1554–1627) remarked in 1591 that he had been introduced to the art of drawing by his father, a biographical detail which underlines if anything the attention paid by Francesco to the education of his son.[52] Don Antonio also inherited his father's passion for alchemy and the chemical arts. It is not clear when exactly Antonio resumed the activities of the alchemical laboratories built by his father but he must have done so quite early because already in 1604 he was aiming to publish the results of experiments in the spagyric arts achieved in the *Casino*.[53] Among the laboratories set up by Francesco I there had also been a glass works, the activities of which were resumed by Antonio sometime before 1601.[54] It

---

[50] Covoni, *Buontalenti ai tempi Medicei*, 18.

[51] "Il Granduca ha ridotto tutte le arti che haveva nel Casino in una sua galeria che ha fatta di nuovo in Palazzo, nella quale Sua Eccellenza sta tutto il giorno quando è in Firenze, et si dice che attende a voler vedere tutte le esperienze che si possano dell'Alchimia," letter by Hercole Conti, Ambassador from Ferrara to Alfonso II (8 Jan, 1583), published in Butters, "Pietra eppure non una pietra," 175. In a letter dating 25 Dec, 1587, to Alfonso II d'Este the Ferrarese Ambassador Hercole Cortile wrote: "Il Granduca [i.e. Ferdinando I de' Medici] ha licentiate tutti quelli che lavoravano nel Casino, et gioellieri et stillatori, dicendo che lui non vuol fare mestiere né di gioie né di stillare," cited by Butters, "Pietra eppure non una pietra," 175. This testimony is confirmed by the following remark by Filippo Pigafetta (1600): "Vieta [Ferdinando] nondimeno à quei ministri [della fonderia], che per niuna maniera diano opera all'alchimia, et alle prove di trasmutar i metalli in oro, ò vero aumentarlo, stimando ciò arte dannosa e del tutto falsa [...]," cited by Butters, "Pietra eppure non una pietra," 145.

[52] For recent research on Antonio Medici's cultural background see Musacchio, *Objects and Identity*. On Francesco and his son: "Il granduca Francesco faceva attendere a quest'arte [del disegno] il marchese D. Antonio suo figliuolo, e ancora adesso seguita non solamente egli, ma tutte le principesse figliuole, e i nipoti del detto granduca Francesco di continuo attendono al disegno ed hanno già messo in istampa qualcosa di loro invenzione [...]," Giovanni Battista Paggi, letter to his brother dated 1591, published in Barocchi, *Scritti d'arte del Cinquecento*, vol. I, 197.

[53] No reference as to when Antonio began to be interested in alchemy can be found in Covoni, *Don Antonio de' Medici*, or in the more recent article by Luti, "Don Antonio de' Medici."

[54] In *L'arte vetraria*, ch. XLII, Neri wrote: "Questo fu il modo che io tenni nel fare la presente Calcidonia l'anno 1601 in Firenze al Casino nella fornace de vetri, nel qual tempo faceva lavorare detta fornace l'egregio Messer Nicolò Landi mio familiare amico e huomo raro nel lavorare di smalto alla lucerna, nella quale fornace feci più padelotti di Calcidonio in detto tempo che sempre venne bello da tutta prova, non uscendo mai delle regole sopra dette e havendo le materie preparate bene."

was on this site and presumably also in the alchemical laboratory of the *Casino* that Antonio Neri (1576–1614), the author of *L'arte vetraria* (1612), made many of his alchemical experiments on the transmutation of colored glass and precious stones.[55] Later, Neri would work glass at both the *Casino* and the glass works set up in Pisa by the alchemist Niccolò Sisti, a former employer of Francesco I and Ferdinando I's chief distiller.

Antonio de' Medici was so interested in alchemy that when he inherited the Casino's library, he added a large number of new books and manuscripts to an already remarkably rich collection. Targioni remarked with contempt that Antonio was infatuated with alchemy and wasted immense quantities of gold on experiments, but at the same time he credited him with putting together a useful collection of pharmaceutical and medical secrets as well as with bringing several arts to perfection.[56] With this artificial comparison, separating the beneficial from the sham and the useless in the *Casino*, Targioni gives us only a vague picture of the breadth and depth of Don Antonio's activities and alchemical interests in the manuscript version of his 1852 published summary. Antonio de' Medici's devotion to alchemy is reflected in his project to publish a book of secrets that would unveil the myriad operations performed in the *Casino di San Marco*'s laboratories.[57] Of this book, which Don Antonio intended to publish in a printing press set up in the same building, only the 1604 frontispiece was printed (Fig. 11). Attached to it are now four manuscript volumes containing some 6,000 alchemical and chemical recipes collected by Don Antonio and his father. The complete publication of this ambitious work may have been interrupted by Neri's travel to Antwerp in 1604 and then never resumed after his return to Florence in 1611.[58] However, an epitome of the collection was published in a small 8° volume with the same title and date; unfortunately, this became so rare that the last trace of it is in a reference in a 1797 note published in *L'osservatore fiorentino*.[59]

The authors of the recipes were both craftsmen employed in the *Casino*, such as Niccolò Landi, and other experts from different parts of Italy and, in a few cases,

[55] Neri announced at the beginning of his work a number of experiments revealing "le pietre che possono trasmutarsi in vetro da quelle che non si possono trasmutare." (Neri, *L'arte vetraria*, 38). On Neri's alchemy, see Grazzini, "Discorso sopra la chimica."

[56] "Si sa che il principe Don Antonio era innamoratissimo, e per meglio dire infatuato dell'Alchimia, e che spese immense somme d'oro per imparare e sperimentare diversi segreti, che gli erano venduti a caro prezzo dagl'impostori, come suol succedere. Peraltro con queste inutili e dispendiose prove, riuscì al principe di raccogliere e verificare un gran numero di segreti appartenenti alla medicina, ed a perfezionare diverse arti: anzi la maggior parte dei preziosi medicamenti, che poi si composero e si dispensarono nella Real Fonderia, ai tempi di Ferdinando II e Cosimo III, erano di quelli acquistati e provati dal principe D. Antonio." (Targioni Tozzetti, *Notizie sulla storia*, 256).

[57] *Apparato,* BNCF, Magl. XVI, 63, I. This printed frontispiece is followed by four manuscript volumes in folio.

[58] This raises the question of Neri's role in the composition of this work. On this see Beretta, *Glass Making Goes Public.*

[59] Third edition (Florence, 1821), Lastri, *L'Osservatore fiorentino*, vol. I, 87.

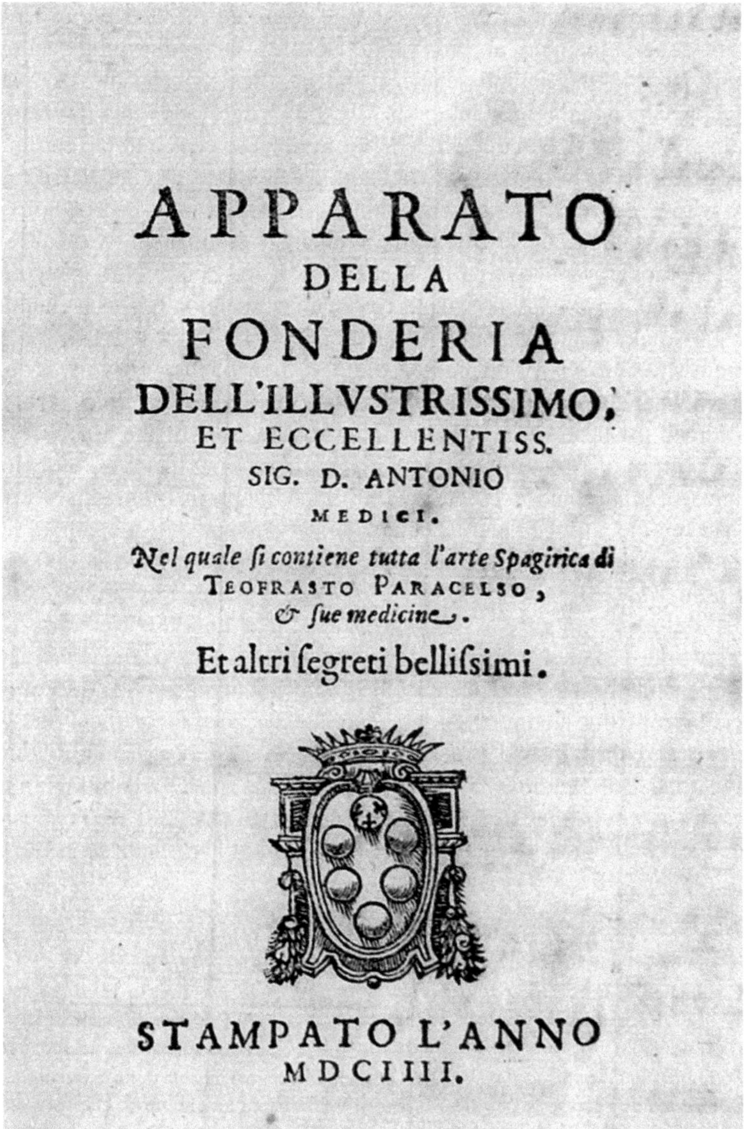

APPARATO
DELLA
FONDERIA
DELL'ILLVSTRISSIMO,
ET ECCELLENTISS.
SIG. D. ANTONIO
MEDICI.
Nel quale ſi contiene tutta l'arte Spagirica di
TEOFRASTO PARACELSO,
& ſue medicine.
Et altri ſegreti belliſsimi.

STAMPATO L'ANNO
MDCIIII.

Fig. 11 Frontispiece of the *Apparato della Fonderia dell'Illustrissimo et Eccellentiss. Sig. D. Antonio Medici* (1604) (Courtesy of Biblioteca Nazionale Centrale, Florence. Mss. Magl. XVI,63/1)

from European countries who sold their secrets to the Medici family. It is likely that among this heterogeneous collection of secrets, many were copied from the manuscripts purchased by Don Antonio.

It would be a mistake to think, as has been customary, of Antonio de' Medici as an isolated and obscure figure in Florence. Quite the contrary, he was an active patron and correspondent of Galileo, and he regarded the *Casino* as a part of a consistent program aimed at promoting alchemy as a discipline endorsing a reformed vision of the natural sciences. Indeed, alchemy and the arts related to it still represented an important body of knowledge, and one far from being undermined with the arrival of Galileo in Florence. The considerable cultural status of these arts helps us to understand the significance of the privileged role assigned to Neri whose expertise in alchemy and glassmaking was prized not only by Antonio de' Medici but also by the nobleman Emmanuel Ximenes (1564–1632).[60] Moreover, Don Antonio's lifelong effort to obtain Neri's secret recipe for the so-called *donum dei*, reveals the intimate connection between chemical arts such as glassmaking and alchemy.

Antonio's *fonderia* was situated on the first floor of the *Casino*, very near the small room where he slept; here, due to a chronic disease towards the end of his life, he was forced to spend most of his time in bed. In front of the *fonderia* hung an imposing portrait of Antonio wearing a black suit bearing the cross of the order of the knights of Malta and decorated with gilded scorpions; these may well have served as a reminder of the pharmaceutical remedy invented by his father Francesco sometime in summer 1580 when he ordered Sisti to capture some 21,000 scorpions for his *fonderia* (Fig. 12).[61] Under the portrait a Latin inscription exalted the chemical skills of Antonio.[62] While we know that Landini, Neri and Giacinto Talducci worked in the laboratory, it is not known who oversaw its activities. Talducci became the superintendent of the *fonderia* only after the death of Antonio and the move of the laboratory, together with the portrait of the Prince, to the *Galleria degli Uffizi*. In 1619, during his stay in Florence, Teodoro Filippo di Liagno (c.1587–1630), known as Filippo Napoletano, made a painting entitled *La bottega del alchimista del Casino* for the Medici (Fig. 13). [63] As is apparent the painting evokes a typical alchemical laboratory and it is difficult to know to what

---

[60] Which is evidenced in the dedicatory epistle of Neri's, *L'arte vetraria.*

[61] Letter published in Barocchi & Bertelà, *Collezionismo mediceo*, 169. In 1576 the numbers of scorpions brought to the *Casino* was 70,000 (Butters, "Pietra eppure non una pietra," 174). Francesco extracted an oil from the scorpions which he used as an ingredient for the *theriaca.*

[62] "Fin dai tempi del Granduca Francesco, tutta una fila di stanze terrene del Casino di San Marco, era stata destinata alla fonderia dei metalli, che il Granduca aveva riempita di molti utensili, di forni di ferro, di barattoli di polvere d'antimonio, e di pietre triturate, ingredienti indispensabili all'uopo. [. . .] Precedeva questa fonderia, una ricca biblioteca, sulla porta della quale Don Antonio teneva appeso un suo ritratto bellissimo, vestito di velluto nero alla spagnola, colla croce bianca di cavaliere [. . .]. In questo il Principe era effigiato qual personaggio fra i più ragguardevoli, ed al piede della figura vedevansi scritti i seguenti versi: *Ingens consiliofactis Antonius ingens // Heic mira insigna quem colit arte locus: // Par Phebo medicas quo vires traxit ab herbis // Aeternum fama unum lumen ab igne tulit.* Questa fonderia, avvenuta la morte di Don Antonio, fu traslocata al pianterreno degli Uffizi, chiamandola la Fonderia dei Granduchi, sulla cui porta si seguitò a tenerci appeso il famoso ritratto." (Covoni, *Don Antonio de' Medici*, 147–8). Elsewhere Covoni makes clear that the 'stanze terrene' were in fact those situated on the first floor.

[63] For more on this painting, see Fornaciai, *Toilette, profumi e belletti*, 56–7; and Lucia Aquino's note in Conticelli, *Alchimia e le arti*, 112–3.

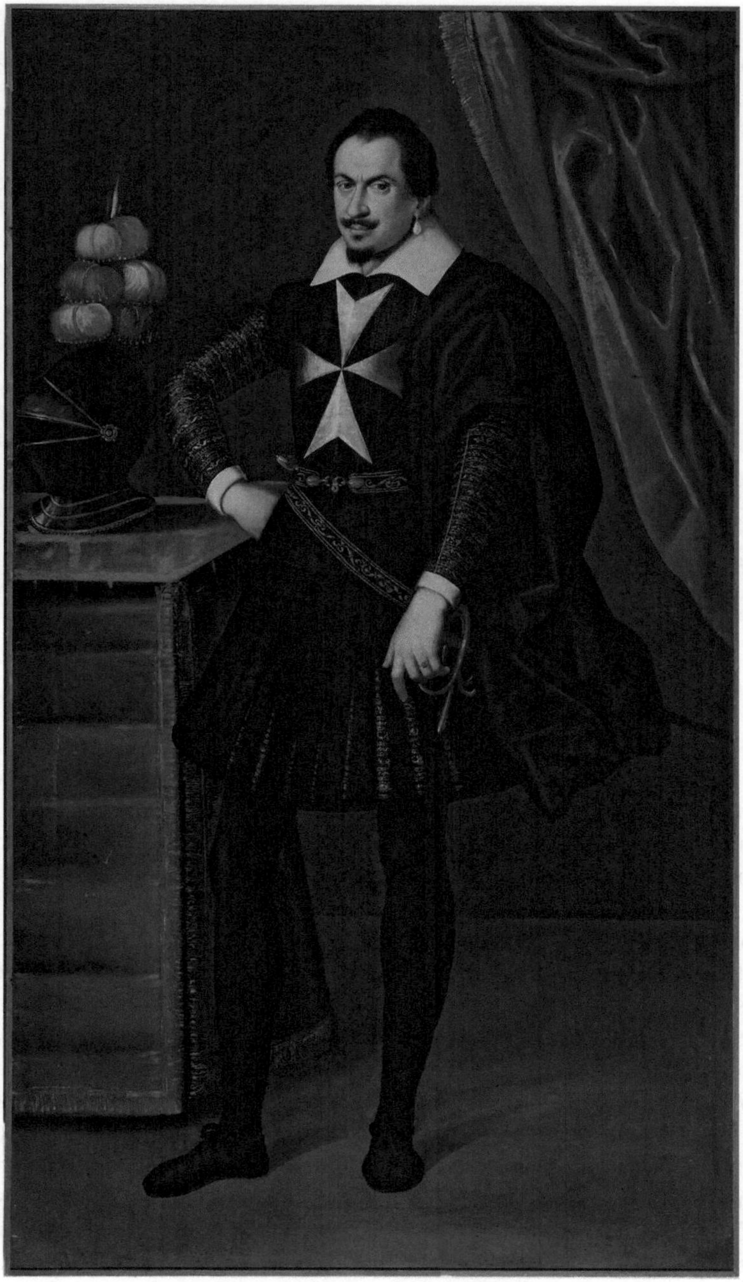

Fig. 12 Portrait of Antonio de Medici. Domenico e Valore Casini (1610–1615) (Courtesy of Galleria degli Uffizi. Florence)

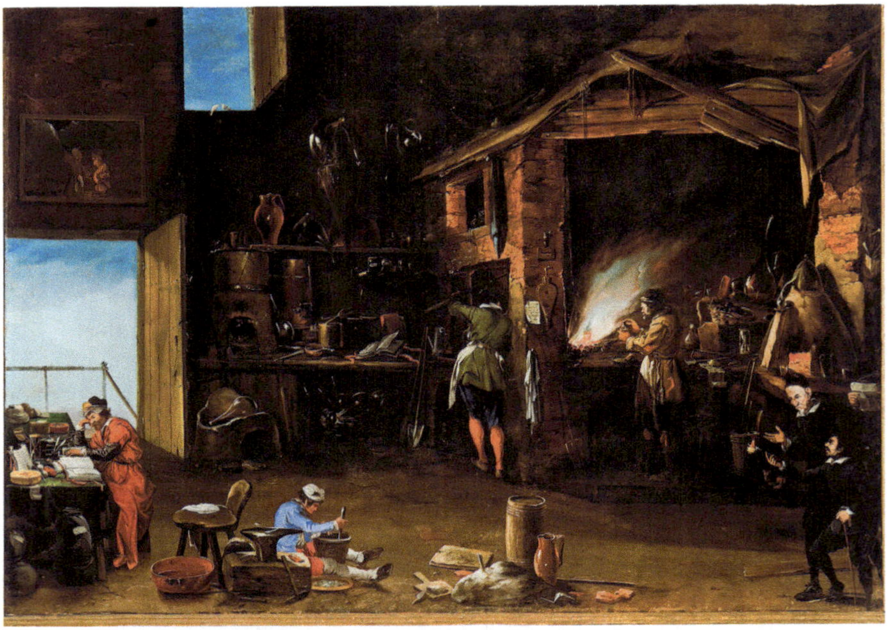

Fig. 13 Teodoro Filippo di Liagno detto Filippo Napoletano. *The Atelier of the Alchemist* (Courtesy of Palazzo Pitti, Florence. Depositi delle Gallerie fiorentine)

extent it depicts Don Antonio's. However, the two figures portrayed on the right suggest that Napolitano was not merely offering an iconographic *topos* of an alchemical laboratory. The small, limping man with a stick looks like Don Antonio: he is holding a luminescent substance probably taken from the furnace, with a nipper in his right hand. Next to Antonio we see an elegantly dressed older man, probably the superintendent of the laboratory, who is looking at the substance held by Antonio. The open door on the left shows the beginning of a descending staircase, suggesting that the laboratory was on the first floor. The rest of Napoletano's workshop offers the usual picture of an alchemical laboratory, with its disorderly arrangement of equipment, its furnace in a poor state, and its dirty floor. Whether authentic or not, Napoletano's representation strongly suggests that by 1619, at the height of Galileo's success, the activities promoted in the *Casino* were still an important feature of Medici scientific patronage. However, in comparison to the Francesco dominated era, the chemical arts seem to have lost their powerful aura and the original strong connections between alchemy, the arts and politics that had been established in the *Casino*. As strongly as Don Antonio believed in the cultural and social importance of alchemy, he was in no position to pursue his father's aim to incorporate these interests into the government of Tuscany. Chemistry, alchemy, pharmacy and medicine remained at the center of Medici patronage, but the new Grand Dukes, by inviting Galileo to court as their philosopher and mathematician, privileged a more pragmatic patronage which

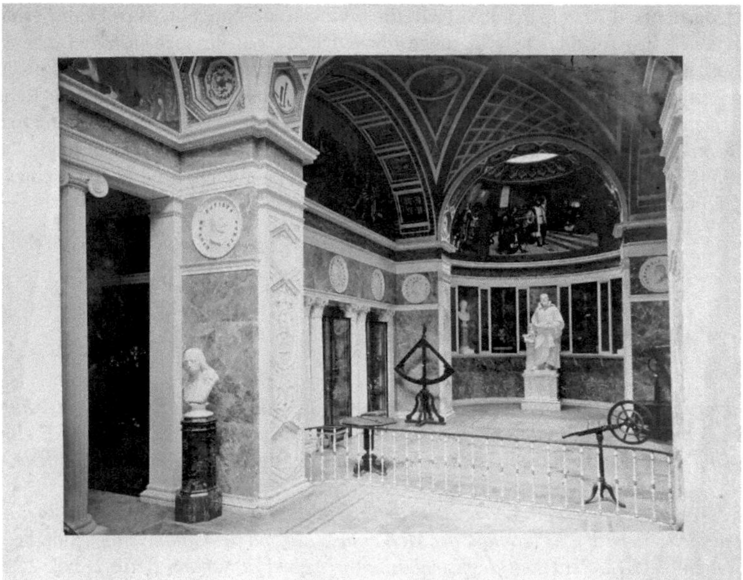

Fig. 14 The Tribuna di Galileo. Photo second half of the nineteenth century (Courtesy of Museo Galileo. Florence)

combined a more encyclopedic view of the ways nature could be investigated. Ironically, however, this new development in Medici scientific-patronage did not have any significant effects on the Tuscan scientific tradition. Although Galileo would soon become a cherished, almost mythical figure, his efforts did not mark a new direction in natural philosophy. The *Accademia del Cimento* adopted Galileo's method only in part and many of its members pursued natural investigations within different traditions, originating in the experiments undertaken in workshops of the *Fonderie* of the *Casino* and the *Uffizi*. At the end of the seventeenth century, important alchemical experiments were still performed with Benedetto Bregans's burning lens at Cosimo III's court and the core of Tuscan science was still dominated by natural history, medicine and chemical arts, as if Galileo had never existed. During the eighteenth century, the *Reale Museo di fisica e storia naturale*, the most important scientific institution founded by the new ruling dynasty of the Hapsburg-Lorraine, privileged the traditional disciplines cultivated by the Medici. In the rooms of the new building, opened to the public in 1775, visitors could admire vestiges of the equipment and minerals used in the *Casino* and the *Uffizi fonderia* in the showcases of the collections on display. It was only much later, in 1841, that the building of the *Tribuna* of Galileo proposed a new hierarchy of scientific display; here telescopes and mathematical instruments occupied the center of a vision of science in which there was no more room for the curious equipment of the past (Fig. 14).

Acknowledgments I wish to thank Suzy Butters, Valentina Conticelli, Sven Dupré, Didier Kahn and Morgan Wesley for their helpful suggestions. Throughout the text *chemical arts* is used to denote those arts regulated by guilds, such as glassmaking, dyeing and pharmacy. More generally, I use the term arts and artists in association with Florentine guilds. Although in Renaissance Florence there is no normative definition for *alchimia* most writers use the term to refer to an art by which it is possible to transmute metals, counterfeit gems and prolong life. Accordingly I apply the same definition to alchemy. It should also be noted that people engaged at the Medici court in alchemical pursuits were called *stillatori*.

Bibliography

1604. *Apparato della Fonderia dell'Illustrissimo et Eccellentissimo Signor Don Antonio. Nel quale si contiene tutta l'Arte spagirica di Teofrasto Paracelso, & sue medicine. E altri segreti bellissimi*. Florence: Biblioteca Nazionale Centrale di Firenze, Magliabecchiana XVI, 63, vols. I–IV.

1980. *La corte il mare i mercati. La rinascita della Scienza. Editoria e Società. Astrologia, magia e alchimia, ed. Firenze e la Toscana dei Medici nell'Europa del Cinquecento*. Florence: Electa.

Albèri, Eugenio. 1842. *Relazioni degli ambasciatori veneti al Senato*, Serie 2, vol. 2. Florence: all'insegna di Clio.

Barocchi, Paola. 1971–1977. *Scritti d'arte del Cinquecento. A cura di Paola Barocchi*, 3 vols. Milan: Ricciardi.

Barocchi, Paola and Giovanna Gaeta Bertelà. 2002. *Collezionismo mediceo e storia artistica*, vol. 1: *Da Cosimo I a Cosimo II, 1540–1621*. Florence: Studio per Edizioni Scelte.

Baroni Vannucci, Alessandra. 1997. *Jan Van Der Straet detto Giovanni Stradano flandrus pictor et inventor*. Milan: Jandi Sapi Editori.

Beretta, Marco. Forthcoming. Glass Making Goes Public: The Cultural Background of Antonio Neri's *L'Arte Vetraria* (1612).

Berti, Luciano. 1967. *Il Principe dello Studiolo. Francesco I dei Medici e la fine del Rinascimento fiorentino*. Florence: Edam.

Bianchi, Maria Luisa, and Maria Letizia Grossi. 1998. Botteghe, economia e spazio urbano. In *Arti Fiorentine. La grande storia dell'artigianato*, vol. 2, eds. Franco Franceschi and Gloria Fossi, 27–63. Florence: Giunti.

Biringuccio, Vannoccio. 1540. *De la pirotechnia*. Venice: per Venturino Roffinello.

Bocchi, Francesco. 1591. *Le bellezze della città di Firenze*. Florence.

Bueno, Mar Rey. 2009. La Mayson pour Distiller des Eaues at El Escorial: Alchemy and Medicine at the Court of Philip II, 1556–1598. *Medical History* 29: 26–39.

Buonsignori, Stefano. 1584. *Nova pulcherrimae civitatis Florentiae topographia accuratissime delineata*. Florence.

Butters, Suzanne B. 1996. *The Triumph of Vulcan: Sculptors' Tools, Porphyry and the Prince in Ducal Florence*, 2 vols. Florence: Leo. S. Olschki.

Butters, Suzanne B. 2000. 'Una pietra eppure non una pietra': Pietre dure e botteghe medicee nella Firenze del Cinquecento. In *Arti Fiorentine. La grande storia dell'artigianato. Il Cinquecento*, vol. 3, eds. Franco Franceschi and Gloria Fossi, 107–185. Florence: Giunti.

Camerota, Filippo. 2008. *I Medici e le scienze*. Florence: Giunti.

Conticelli, Valentina. 2006. Lo Studiolo di Francesco I e l'alchimia: nuovo contributi storici e iconologici, con un carteggio in appendice (1563–1581). In *L'Art de la Renaissance entre science et magie*, ed. Philppe Morel, 207–268. Paris: Somgy.

Conticelli, Valentina. 2007. *'Guardaroba di cose rare et preziose': Lo Studiolo di Francesco I de' Medici. Arte, storia e significati*. Lugano: Agorà.

Conticelli, Valentina. 2012. *L'Alchimia e le arti. La fonderia degli Uffizi da laboratorio a stanza delle meraviglie*. Sillabe: Livorno.

Corazzini, Giuseppe Odoardo. 1900. *Diario Fiorentino di Agostino Lapini dal 252 al 1596*. Florence: Sansoni.

Covoni, Pierfilippo. 1892a. *Don Antonio de' Medici al Casino di San Marco*. Florence: Tipografia cooperativa.

Covoni, Pierfilippo. 1892b. *Il Casino di San Marco costruito dal Buontalenti ai tempi Medicei*. Florence: Tipografia cooperativa.

Del Riccio, Agostino. 1996. *Istoria delle pietre*. Torino: Allemandi.

Doren, Alfred. 1940. *Le Arti fiorentine*, 2 vols. Florence: Le Monnier.

Fara, Amelio. 1988. *Bernardo Buontalenti. L'Architettura, la guerra e l'elemento geometrico*. Genova: Sagep.

Fornaciai, Valentina. 2007. *Toilette, profumi e belletti alla corte di Medici. Un itinerario fra i ricettari di corte, le collezioni fiorentine e gli ambienti di residenza medicea*. Livorno: Sillabe.

Franceschi, Franco, and Gloria Fossi. 1999. *Arti Fiorentine. La grande storia dell'artigianato*, vol. 2. Florence: Giunti.

Galluzzi, Ruguccio. 1781. *Istoria del granducato di Toscana sotto il governo della casa Medici*, 8 vols. Livorno: da Giovani-Tommaso Masi e compagni.

Galluzzi, Paolo. 1982. Motivi paracelsiani nella Toscana di Cosimo e di Don Antonio dei Medici: alchimia, medicina chimica e riforma del sapere. In *Scienze, credenze occulte, livelli di cultura*, ed. Giancarlo Garfagnini, 31–62. Florence: Leo S. Olschki.

Goldthwaite, Richard. 1980. *The Building of Renaissance Florence*. Baltimore: Johns Hopkins University Press.

Goldthwaite, Richard. 1999. Realtà economico-sociale e status culturale dell'artigiano. In *Arti Fiorentine. La grande storia dell'artigianato*, vol. 2, eds. Franco Franceschi and Gloria Fossi, 14–18. Florence: Giunti.

Grazzini, Maria Grazia. 2012. Discorso sopra la Chimica: The Paracelsian Philosophy of Antonio Neri. *Nuncius* 27: 411–467.

Heikamp, Detlef. 1970. L'antica sistemazione degli strumenti scientifici nelle collezioni fiorentine. *Antichità viva* 9: 3–25.

Heikamp, Detlef. 1986. *Studien zur mediceischen Glaskunst: Archivalien, Entwurfszeichnungen, Gläser und Scherben*. Florence: Kunsthistorisches Institut.

Ilardi, Vincent. 2007. *Renaissance Vision from Spectacles to Telescopes*. Philadelphia: American Philosophical Society.

Lastri, Marco Antonio. 1821. *L'Osservatore fiorentino sugli edifizi della sua patria*. Florence.

Luti, Filippo. 2008. Don Antonio de' Medici professore de' secreti. *Medicea* 1: 34–47.

Miniati, Mara. 2000. 'Un fabbro sia un buon maestro'. La produzione di strumenti scientifici a Firenze nel Cinquecento. In *Arti Fiorentine. La grande storia dell'artigianato. Il Cinquecento*, vol. 3, eds. Franco Franceschi and Gloria Fossi, 263–295. Florence: Giunti.

Musacchio, Jacqueline M. 2007. Objects and Identity. Antonio de' Medici and the Casino at San Marco in Florence. In *The Renaissance World*, ed. John Jeffries Martin, 481–500. London: Routledge.

Neri, Antonio. [1612] 2001. *L'arte vetraria*, ed. Ferdinando Abbri. Florence: Giunti.

Parigino, Giuseppe Vittorio. 1999. *Il Tesoro del Principe. Funzione pubblica e privata del patrimonio della famiglia Medici nel Cinquecento*. Florence: Leo S. Olschki.

Perifano, Alfredo. 1997. *L'alchimie à la Cour de Côme Ier de Médicis: savoirs, culture et politique*. Paris: Honoré Champion.

Piccardi, Giovanni. 2005. La Fonderia Medicea di Firenze. In *Atti del XI Convegno Nazionale di Storia e Fondamenti della Chimica*, eds. Luigi Cerruti and Francesca Turco, 197–210. Roma: Accademia delle Scienze detta dei XL.

Shaw, James, and Evelyn Welch. 2011. *Making and Marketing Medicine in Renaissance Florence*. Amsterdam: Rodopi.

Staley, Edgcumbe. 1906. *The Guilds of Florence*. London: Methuen.

Targioni Tozzetti, Giovanni. 1769–1779. *Selve di notizie, spettanti all'origine de' progressi e miglioramenti delle scienze fisiche in Toscani, messe insieme dal Dottor Giovanni Targioni-Tozzetti, per uso del dottor Ottaviano suo figlio*, 17 vols. Biblioteca Nazionale Centrale di Firenze, Targioni. Tozzetti, Ms. 189.

Targioni Tozzetti, Giovanni. 1780. *Notizie degli aggrandimenti delle scienze fisiche accaduti in Toscana nel corso di anni LX. del secolo XVII*, 3 vols. Florence: G. Bouchard.

Targioni Tozzetti, Giovanni. 1852. *Notizie sulla storia delle scienze fisiche in Toscana cavate da un manoscritto inedito*. Florence: Biblioteca Palatina.

Tosi, Alessandro. 1989. *Ulisse Aldrovandi e la Toscana. Carteggio e Testimonianze documentarie*. Florence: Leo S. Olschki.

Vossilla, Francesco. 1993. Cosimo I, lo scrittoio del Bachiacca, una carcassa di capodoglio e la filosofia naturale. *Mitteilungen des Kunsthistorischen Institutes in Florenz* 37: 381–395.

Goldsmiths and Chymists: The Activity of Artisans Within Alchemical Circles

Lawrence M. Principe

Abstract The study of early modern alchemical practitioners has recently expanded from the work and writings of natural philosophers to include the labors of artisans. Yet there remains relatively little clear documentation about the specifics of such artisanal chymical practices or about the degree of significant epistemic exchange that may have taken place between artisans and better-known chymists. This paper examines several gold- and silversmiths who were part of an extended alchemical network that stretched across the Netherlands, France, and England in the mid-seventeenth century. The silversmith Anthoni Grill (and to a lesser extent his brother Andries) worked on refining and transmutational processes with impressive chymical expertise and eventually achieved international notoriety and social and financial success. Grill built large laboratories, managed laborants, engaged in collaborations, mined texts for information, and shared results. Several Parisian goldsmiths were likewise tied into this chymical network where they exchanged experimental and theoretical information with such figures as Kenelm Digby, Samuel Cottereau Duclos, and Johann Rudolf Glauber. In England, George Starkey found that the expertise of gold- and silversmiths could, alternatively, be inconvenient when trying to sell alchemical metals. The documentation provided here blurs the boundary between artisans and natural philosophers, at least in chymistry.

Alchemy, or to speak more inclusively and less anachronistically, *chymistry* was a familiar topic in early modern Europe. Despite its well-established culture of secrecy, it was not only highly visible and widely-known, but widely-practiced across both geographical and social boundaries. Indeed, present-day scholars continue to explore the multifaceted domains of alchemy, revealing ever more clearly

L.M. Principe (✉)
Department of History of Science and Technology, Johns Hopkins University, Baltimore, MD, USA

Department of Chemistry, Johns Hopkins University, Baltimore, MD, USA
e-mail: Lmafp@jhu.edu

S. Dupré (ed.), *Laboratories of Art*, Archimedes 37,
DOI 10.1007/978-3-319-05065-2_7, © Springer International Publishing Switzerland 2014

and emphatically its broad and pervasive extent and the important role it played in early modern European culture.[1]

Some years ago, I pointed to the significant diversity that existed within the alchemical tradition of early modern Europe.[2] Contrary to what many alchemical authors themselves would have us believe of their art, and what much of the earlier secondary literature claimed, the sages did not "all say one thing." Moving beyond the rhetorical representations of alchemy as a largely monolithic and static tradition, scholars now recognize that early modern chymical thought was both diverse and dynamic. Vigorous disagreements and successive innovations characterized virtually its entire history, and most of all, its early modern existence. On the broadest scale, opinions varied in regard to what chymistry's main goals should include—*chrysopoeia* (metallic transmutation), *chemiatria* (medicinal applications), commercial production, and so on. Even within any one of these important branches, both theoretical frameworks and practical approaches varied widely. What was the correct hidden composition of metals, and of matter more generally? What was the correct starting material for the Philosophers' Stone and how should it be processed? Which authorities were trustworthy and which should be rejected? Theoretical choices and commitments both guided and were guided by practical experimentation in the laboratory, leading to new ideas and practices as workers reinterpreted older authorities to fit their observations or struck out in new directions to achieve a variety of goals.[3] Thus diversity is to be found not only in theoretical notions but in practices as well; the dynamic interaction between head and hand stands as a hallmark of the chymical tradition.

Yet another aspect of chymical diversity lies with the practitioners themselves. Who pursued and practiced early modern chymistry? Naturally enough, modern scholarly inquiry focussed first on those who contributed to the enormous flood of early modern chymical publications, since those sources were the most obvious and most readily available. Further inquiries began to explore the vast bulk of manuscript and archival materials, with the result that the ambit of "the alchemist" has been broadened to include practitioners representing the broadest possible range of early modern people.[4] University-educated physicians, natural philosophers, and scholars busied themselves with chymistry, as did a range of empirics, entrepreneurs, and enthusiasts, as well as an assortment of artisans that included artists, metalworkers, brewers, and cobblers. Chymical discourse and practice took place in academic chambers, in courtly settings, in private homes, in commercial

[1] For an overview of the history of alchemy, see Principe, *Secrets of Alchemy*. On the use of the term 'chymistry'. see 84–5. For a study of the words chemistry and alchemy and their changing referents, see Newman & Principe, "Alchemy vs. Chemistry."

[2] Principe, "Diversity in Alchemy," and "Robert Boyle and Isaac Newton."

[3] On the interplay of theory and practice in alchemy, see Newman & Principe, *Alchemy Tried in the Fire*, esp. 92–206.

[4] For example, see Nummedal, *Alchemy and Authority*, which details the roles and status of 'entrepreneurial' chymists in German courts. For a sense of the broad sweep of early modern alchemical practice, see Moran, *Distilling Knowledge*.

establishments, and in noisy workshops. Most recently there has been particular interest in exploring alchemy's place among artists and artisans, an investigation that rightly reflects the crucial position of material production within chymistry throughout its history. Important questions remain however in regard to demonstrating the specific theoretical and practical content of such artisanal endeavors, and in terms of identifying connections between them and the better documented and recognized work carried out in more visible and more intellectually elevated circles.[5] How did artists and artisans understand chymistry, and to what extent did they practice it? To what extent were they familiar with more bookish or scholarly ideas, practices, practitioners (and vice versa)? Can exchanges of knowledge between the two groups be clearly documented? Can we identify what practices, materials, or ideas passed between them, and how one group might have responded to information from the other? Answering such questions promises to further illuminate chymistry's cultural and intellectual place in early modern Europe.

Some such connections seem relatively clear. Chymical workshops appear frequently in Netherlandish genre painting, indicating that Northern artists were clearly aware of the figure of the alchemist and his iconic value.[6] Likewise, the allegorical images of widely varying artistic quality in chymical books and manuscripts imply some sort of contact between author/practitioner and artist. Yet such artworks do not necessarily indicate that the artists were actually in direct and meaningful contact with chymists. On a purely commercial level, artists would have been in contact (possibly directly in some instances, but probably more frequently indirectly through a retailer) with makers of chymically-prepared pigments such as vermilion or white lead, just as metalworkers and assayers would have connected with producers of mineral acids, fabric-makers with dye manufacturers, and so forth. Yet none of these contacts tells us anything about possible epistemic exchanges or collaborations on the level of practical or theoretical knowledge.

Documented examples of extended collaborations or exchanges between artisans and natural philosophers would be both more interesting and more informative for mapping out the domains and dynamics of early modern alchemical practice. A promising locus for such study is to be found particularly among gold- and silversmiths. On the one hand, their trade regularly involved chymical processes such as assaying, refining, and so forth, and they were thus more or less conversant with chymical substances (acids, salts, metals, amalgams, etc.) and chymical operations (fusion, quartation, distillation, etc.) and would frequently have maintained workshops equipped with the necessary equipment (furnaces, flues, fuel, retorts, crucibles, etc.) needed for chymical work generally. On the other hand, the basic materials of their artisanal labors—gold and silver—held a central

[5] See Smith, *Body of the Artisan*.

[6] Principe & DeWitt, *Transmutations*; and Brinkman, *Alchemist in de prentkunst*. Lennep, *Art et alchimie*, is also useful, especially for its illustrations, but suffers from an inaccurate understanding of alchemy and its practice.

position in chymistry, that is to say, in the transmutational endeavors that figured so prominently in early modern chymistry and that promised such highly profitable returns. Thus, the artisanal experience of gold- and silversmithing provided a certain commonality with the pursuit of metallic chymistry, and might easily have inclined such artisans toward transmutational endeavors, and hence toward contact with other aspiring transmuters who were not themselves artisans. The practical experience of gold- and silversmiths also rendered them sources of information for those wishing to know more about the properties and potentialities of the precious metals. Accordingly, gold- and silversmiths figure prominently in many examples of the genre of "transmutation histories"—detailed published accounts of successful transmutations—where such artisans are routinely called in to assay a sample of the gold or silver produced, and thus act as expert witnesses to its authenticity.[7]

Fortunately for such a study, there existed a mid seventeenth-century network of alchemical practitioners where such interactions can be well documented. From the 1640s through the 1660s, a network of correspondents and collaborators that included several gold- and silversmiths linked three major urban centers—Amsterdam, Paris, and London—and freely crossed political, linguistic, social, and confessional boundaries. Its numerous participants shared ideas, experiences, and aspirations relating to a variety of chymically-based projects, and, crucially for such a project as this one, this network left behind a wealth of written records that provide historians with the necessary clear evidence from which to work. I will therefore focus here on the chymical activities of gold- and silversmiths within this group and their interactions with its other members and with the broader realms of chymistry.

Anthoni and Andries Grill in the Netherlands and Sweden

One part of this network has already been extensively studied, namely the portion gathered around the intelligencer Samuel Hartlib (1600–1662) in London.[8] The Hartlib Circle's utopian, educational, and commercial schemes made chymistry a subject of particular interest due to its lucrative potential. An important contact in this regard was the former Calvinist minister Johann Moriaen (c.1591–1668) in Amsterdam. Moriaen both pursued his own chymical interests and aspirations and functioned as a communications hub for others who shared those interests. Working from Moriaen's surviving correspondence with Hartlib and his protégé Benjamin Worsley (1618–1673), John Young has beautifully documented several chymical

[7] On transmutation histories, see Principe, *Secrets of Alchemy*, 167–70, *Aspiring Adept*, 93–8, 108–11; and Newman, *Gehennical Fire*, 3–13.

[8] Classic treatments of the Hartlib Circle include Turnbull, *Hartlib*; and Webster, *Great Instauration*.

projects in which Moriaen was involved. One of these projects, undertaken in 1651, involved trying to produce silver from tin. Four hopeful experimenters collaborated on this project: Moriaen, Worsley, the German inventor and chymist Johannes Küffler (1595–1677), and one whom Young calls a "very shadowy figure" named Anthoni Grill and who is most often designated simply as *Aurifaber*, that is, the goldsmith.[9]

Grill is, in fact, not so shadowy a figure. He has left behind substantial historical traces, artifacts, and documentation. Furthermore, while his involvement with transmutational chymistry has hitherto been taken as a failure, fresh research now not only documents his laudible chymical ingenuity and engagement with contemporaneous scholarship, but also reveals that he eventually turned his chymical expertise into an enormous success. Grill thus provides an outstanding and detailed example of the alchemical activity of an artisan.

Anthoni Grill was born about 1607 in Augsburg. He and two brothers, all silver- and goldsmiths like their father Balthasar Grill before them, left Germany to practice their trade in the Netherlands.[10] Several pieces of fine silverwork by Anthoni's younger brother Johannes (c.1614–1670) survive, as do a few pieces probably by Anthoni himself (Figs. 1 and 2).[11] Their elder brother Andries (1604–1665), who set up shop in The Hague, is considered by some to have been one of the finest gold- and silversmiths in the mid-century Netherlands and several pieces of his work also survive (Fig. 3).[12] Anthoni arrived in Amsterdam around 1630, married in 1634, and became established as a silversmith on the Koningsstraat. Yet his interests and ambitions soon led him beyond the artisanal working of precious metals. Notarial records indicate that already in 1635 he undertook a project to reclaim silver, copper, and lead from used *testen* (cupels), indicating an interest, and presumably some degree of technical ability, in chymical processes. By 1649 he was known as an assayer as well as a silver- and goldsmith. The notarial records also report that in 1651, he paid the handsome sum of 12,000 guilders (over £1,000) for a house with a large yard on the Looiersgracht in order to set up a *smelterij* (smelter) for obtaining gold and silver.[13]

This Dutch archival record is corroborated and elaborated in letters from Moriaen in Amsterdam to Worsley in London. These letters also recount the extent of Grill's devotion, growing knowledge, and labor in chymical processes. Moriaen

---

[9] Young, *Johann Moriaen*, 226–9, 231. Young translates *Aurifaber* as *goldmaker* (which would have been *aurifactor*) rather than as *goldsmith*. The identification of *Aurifaber* as Grill is made in Hartlib's *Ephemerides* for 1650, *Hartlib Papers* (hereinafter HP) 28/1/49B. For Worsley, see Leng, *Benjamin Worsley*. On Küffler, the son-in-law of the inventor Cornelius Drebbel, see Young, *Johann Moriaen*, 52–7; and Principe, *Aspiring Adept*, 85–6.

[10] Eeghen, "Grill's Hofje."

[11] Lorm, *Amsterdams goud en zilver*, 44–7, 54–6, 59–60; brief biographies on 506–7 by Dirk Jan Biemond. I am grateful to Dr. Lorenz Seelig for this reference, and for his generous and invaluable comments upon this paper.

[12] Gelder, "Haagsche bekerschroef."

[13] Eeghen, "Grill's Hofje," 50.

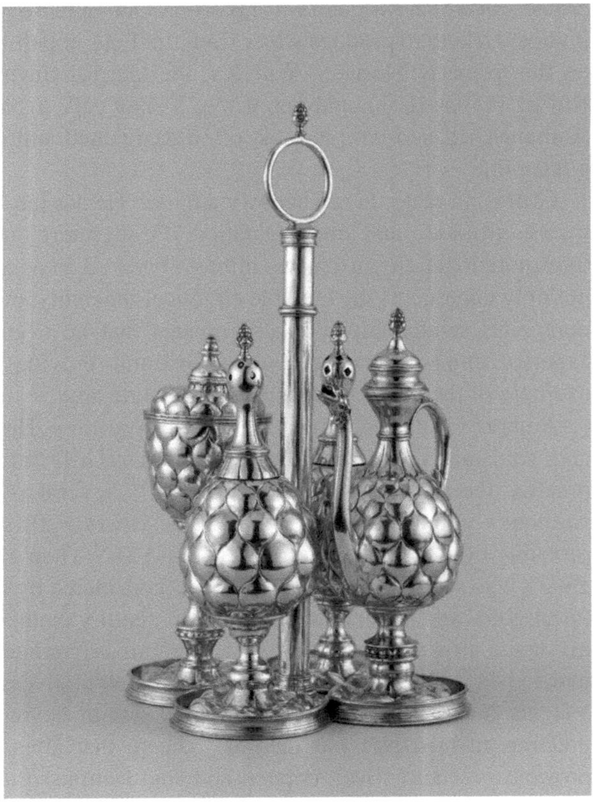

Fig. 1 Silver and part-gilt cruet and spice set attributed to Anthoni Grill, 1642 (Courtesy of Rijksmuseum Amsterdam, BK–1997–1)

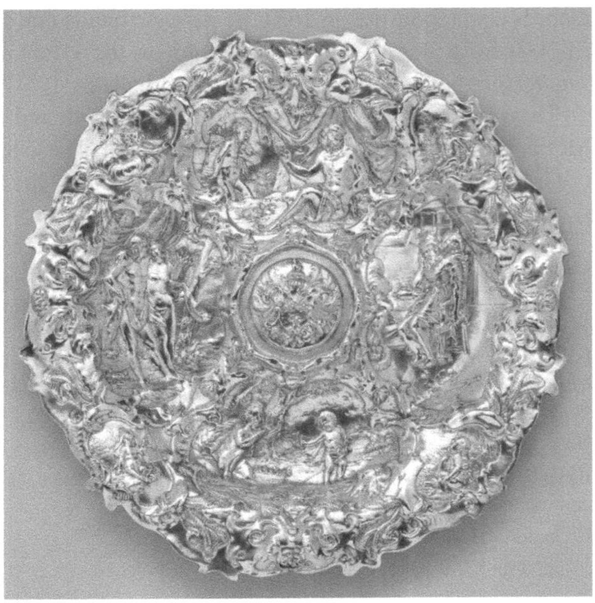

Fig. 2 Silver platter showing allegories of the four seasons, four ages of man, and four elements by Johannes Grill, 1649 (Courtesy of Rijksmuseum Amsterdam)

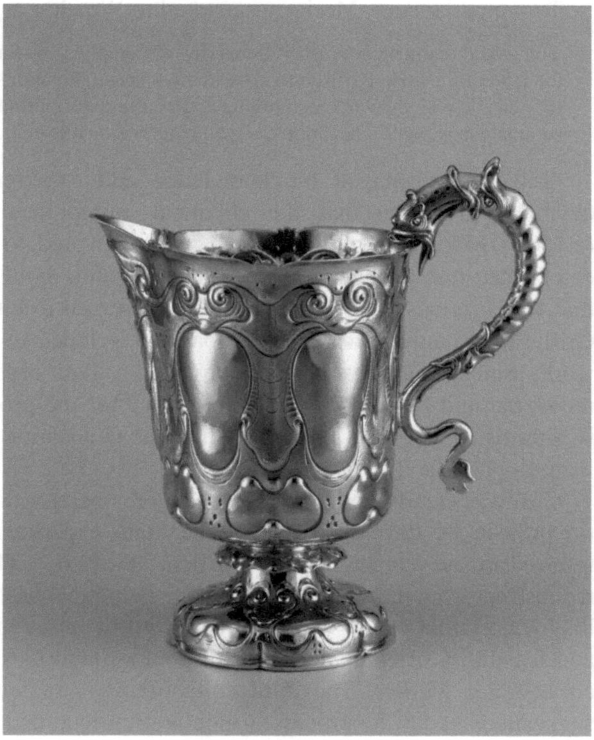

Fig. 3 Silver pitcher by Andries Grill, 1649 (Courtesy of Rijksmuseum Amsterdam, BK–NM–13270–B)

reports that Grill's decision to purchase the property on the Looiersgracht arose from their collaborative project for extracting silver from tin for which Moriaen's own furnace was found to be "too narrow and feeble." Yet Grill's aspirations went much further than this one project, for he set up on the property, besides his lodgings, six laboratories [*laboratoria*]

> not so much indeed for our work as for his own, this same work, I say, which he has discovered with his own thoughts from reading partly Paracelsus and partly Glauber [. . .]. He knows besides how to draw gold or silver out of any talc, just as the talc is disposed and endowed by nature. This knowledge indeed spawned such boldness that he did not hesitate to spend 25,000 guilders or more in buying the estate and in erecting things necessary for the work, and he is sparing no labor in completing them as quickly as possible, and is leaving no stone unturned.[14]

[14] "Aurifaber, non procul ab aedibus meis 12 florenorum millibus fundum sibi coëmit, in quo praeter domicilium sex laboratoria diversa exstruit, jam de opere certus non tam nostro, quam suo ipsius, isto inquam opere quod ex lectione, partim Paracelsi partim Glauberi, propriâ meditatione consequutus est [. . .] novit praeterea ex quolibet Talco argentum vel aurum elicere prout Talcum dispositum et ditatum a Natura est. Haec porrò scientia peperit audaciam illam ut non dubitaret 25 florenorum millia aut amplius in emendo fundo et exstruendis ad opus necessariis, impendere in quo etiam quàm citissimè perficiendo nulli labori parcit, nullum non movet lapidem [. . .]." (Moriaen to Worsley, 31 March 1651, HP 9/16/3A–4B, on 4A).

Two months later, Moriaen remarked to Worsley:

> You would really be astonished if you saw the supplies, instruments, and beautiful order of the laboratories! He [Grill] and his wife now spend the nights in the new lodgings so that they can be with the workers morning and evening; the building is progressing quickly, but so much construction nevertheless requires a reasonable amount of time.[15]

Grill also displayed his knowledge and experience by critiquing several chymical processes that the chymical entrepreneur Johann Rudolf Glauber (c.1604–1670) had boasted about and endeavored to sell.[16] In particular, Grill determined that Glauber's process for separating gold and silver "does not excel that which is customary to him [Grill] in regard to either the labor or the cost." Grill's own method, which he probably developed in the course of his assaying work, used special vessels (made of gold!) to limit breakage and loss and cleverly recovered and recycled the acid necessary for the process, thus cutting costs by two-thirds. Hence, as Moriaen asserts, "the Goldsmith's method exceeds others in its utility and facility."[17]

Grill was clearly much more than a mere operator. In terms of the collaborative tin project, he developed one of the main experimental protocols to be used, referred to by Worsley as the "Grillian" method.[18] Though it is not possible to reconstruct from the sources at hand the theoretical basis upon which he developed his method, this process was characterized by an attention to operational facility and feasibility in comparison with the more laborious method proposed by their collaborator Küffler. The third collaborator, Worsley, also offered operational suggestions about the process, which are not recorded, but from other contemporaneous sources, Worsley seems to have had more enthusiasm about the potential of chymistry than experience in its practice.[19] Moriaen reported that Grill's workers

---

[15] "[...] mirareris profecto si Laboratoriorum copiam, apparatum et ordinem decentem videres. Iam ipse cum Uxore in novo hospitio pernoctant, ut laborantibus matutinè et sero adesse possint, aedificium etiam celeriter progreditur sed tot ac tanta, moram tamen requirunt justam." (Moriaen to Worsley, 26 May 1651, HP 9/16/6A).

[16] On Glauber, see Young, *Johann Moriaen*, 183–98, *passim*; and Smith, "Vital Spirits."

[17] "Illa de qua Glauberus gloriari solebat, aurifabro judice, non antecellit hanc sibi usitatam, vel quò ad laborem vel quò ad lucrum [...]. Aurifabri artificium aliis et utilitate et facilitate suâ praestat." (Moriaen to Worsley, 31 March 1651, HP 9/16/3A). Grill's method isolated the residual silver nitrate from the parting, and, rather than recovering the silver by the more usual method of precipitation with copper, instead destructively distilled the nitrate, leaving the silver as a residue and allowing much of the nitric acid used initially to be recovered as a distillate. Grill also took the unusual step of using vessels made of gold ["vasis suis aureis"], rather than of glass, for distilling the acid. Despite the cost of the precious metal, its durability (and inertness) relative to constantly breaking or exploding glass vessels, actually saved money, considering that there would be no loss of the precious metal. Hartlib might be referring either to Grill or to a member of his family still in Augsburg when he recorded that Moriaen had informed him that "the Goldsmith at Augsb[urg] made his melting or refining Pots instead of glass or earth of pure gold wherby hee hath gained or saved many thou[sand]s," see HP 28/2/17B.

[18] Moriaen to Worsley, 26 May 1651, HP 9/16/6A–6B.

[19] Newman & Principe, *Alchemy Tried in the Fire*, 236–56.

could consistently prepare two and one-half pounds (five Marks) of silver from 100 pounds of tin, a quantity augmentable by repeating the process on the untransmuted tin recovered. It is possible that they were separating silver present as an impurity in the tin and/or the lead used in the process, although Moriaen's claim to have eventually turned 27 % of the weight of the tin into silver is implausible. In the end, the project for obtaining silver from tin seems not to have fulfilled initial expectations given Moriaen's and Grill's subsequent financial difficulties, even though (or perhaps because) Grill, according to one report, spent £1,200 on it.[20]

Nevertheless, some initial success may stand behind the contract that Grill signed in July 1653 with his brother Andries to supply him with silver, presumably for use and sale in his silversmithing business.[21] Significantly, Andries himself worked on transmutational alchemy as well. The Danish traveller, savant, and chymical author Olaus Borrichius (1626–1690) was told during his trip through the Netherlands in 1662 that "the goldsmith Grill of The Hague, a German, drew a great deal of lead, some silver, and some gold, from Norwegian lead ore." The ore came from Skien, and Andries worked on it in hundred pound quantities, implying that his operational resources were substantial.[22] Rather more interesting is a second report Borrichius obtained in Paris three years later in 1665:

> Grill of The Hague [...] dissolved or cooked a piece of lead in spirit of salt for several months, saw something star-like in the midst of the spirit [...] and after the liquid had evaporated, by assaying the remaining material in a cupel, he found a sixth part of it to be good silver, but afterwards was unable to find a similar spirit of salt.[23]

At least as surprising as the transmutation itself is the fact that Andries Grill did this experiment "upon the advice of Bohn of Leipzig," surely the famous chymist, author, and professor of medicine at the University of Leipzig, Johannes Bohn (1640–1718). Under what circumstances would a silversmith in The Hague be in contact with a professor of Leipzig? Since the experiment lasted several months, it must have been begun in 1664, well before Bohn had published anything on chemistry, and even before he formally received his doctorate. In 1663–1664, however, Bohn was travelling through Northern Europe, including the Netherlands, and thus must have met and spoken with Grill at that time, and suggested this transmutational experiment to him. This solution begs the question of why Bohn

---

[20] HP 9/16/3B; "hee [Grill] adventur's 12. hundred lb upon an Exp[eriment] of Tinne and something else in which Mr. Mor[iaen] hath also an Adventure and is a very promising busines." (Hartlib, *Ephemerides* 1651, HP 28/2/15A). This figure might simply refer to the purchase of the estate, rather than to the direct costs for the tin project.

[21] Eeghens, "Grill's Hofje," 50. Much more about Anthoni's activities and aspirations could certainly be gained from a closer inspection of the many acts recorded by his notary Justus van de Ven, such as this contract with Andries, which I have not been able to examine directly.

[22] Borrichius, *Itinerarium*, vol. II, 58–9, 1 February 1662.

[23] Borrichius, *Itinerarium*, vol. IV, 276, 4 March 1665.

would have visited Grill in the first place, unless Grill had achieved some notoriety for chymical interests or knowledge, that is, in topics of interest to Bohn as well.

Whatever notoriety Andries Grill had obtained prior to Bohn's visit was substantially augmented shortly thereafter when one of the most celebrated reports of the Philosophers' Stone recounted the same experiment with lead that Borrichius had recorded privately. In late 1666, Johann Friedrich Helvetius (1625–1709), physician to the Prince of Orange (1650–1702), reportedly received a visit at his house in The Hague from a mysterious stranger. This visitor showed Helvetius some heavy lumps of material that he claimed to be the Philosophers' Stone that he had prepared, and eventually gave Helvetius a tiny sample. After the visitor departed, Helvetius melted some lead, cast in the crumb of material according to the instructions he had been given, and thereby produced pure gold that was afterwards successfully examined by a goldsmith and then formally assayed by a local silversmith named Brechtel, an artisan undoubtedly identifiable as Hans Coenraet Brechtel (1609–1675), a native of Nuremberg, resident in The Hague from 1640, and the most prominent silversmith in the city at this time.[24] Helvetius published this account as *Vitulus aureus* in 1667, and the story gained wide renown across Europe, provoking the interest of many, including Benedict Spinoza (1632–1677) and Robert Boyle (1627–1691).

As a preamble to his own account, Helvetius reported other transmutational experiences, one of which "was done at The Hague by a certain silversmith whose name is Grill."[25] Helvetius recounts that this "silversmith and student of alchemy much-engaged in the art" obtained some spirit of salt "not prepared by the common method" from Helvetius's friend Johann Caspar Knöttner, a cloth dyer (and apparently, another type of artisan employed in chymical work). Consistent with Borrichius's private diary's account of the event, Helvetius tells how Grill poured this spirit of salt over some lead in a glazed vessel ("such as is used for preserves and confections") and after two weeks was delighted to see a silvery "star" floating in the liquid—perhaps some sort of crystalline concretion. Grill told Helvetius of this phenomenon, which the silversmith believed to be the "signate star of the philosophers that he had read about in Basilius [Valentinus]." Helvetius and "many other honest men" went to see the experiment and marveled at it. After the summer's heat had evaporated the liquid, Helvetius continues, Grill took some of the now spongy and ashen-colored lead, with a piece of the star-like formation, and cupelled it. From one pound he obtained 12 ounces of silver, and out of this silver he isolated two ounces of gold. Helvetius notes that he still owns some of Grill's spongy lead, a piece of the star, and samples of the transmuted silver and gold, all of which he can display to interested or as yet unconvinced inquirers.

Unfortunately, this wonderful success did not make Andries rich. For he tried to learn Knöttner's method for preparing the spirit of salt, but the cloth-dyer had in the

---

[24] For Brechtel (or Breghtel) and his work, see Pijzel-Dommisse, *Haags goud en zilver*, 30: "zonder twijfel de belangrijkste en meest veelzijdige Haagse zilversmid uit de zeventiende eeuw."

[25] Helvetius, *Vitulus aureus*, 831–2.

meantime forgotten exactly which sort of spirit of salt he had given Grill, and while trying to determine which it was, "he and his entire family were seized by the plague, and died." As for the silversmith, shortly thereafter "he fell into the water and exchanged life for death; no investigator after the death of these two has been able to discover the art of making it."[26]

Returning again to the younger brother Anthoni in Amsterdam, we find that his chymical work and interests continued unabated despite the collapse of the collaborative venture on tin. A 1657 letter to Moriaen indicates that by that time Grill had travelled to the coppermines of Sweden and done assaying there. His account of Swedish ores and answers to chymical questions posed to him (probably from Worsley) display significant knowledge of metallic smelting, refining, and assaying. The same letter also reports that he had just purchased the astounding quantity of *fourteen thousand* pounds of English or Scottish *potloet*, that is, the lead compound—mostly litharge (lead oxides)—used for glazing earthenware.[27] Grill complained however, that he was cheated in some unspecified way, such that the material proved unsatisfactory for his intended purpose. He nevertheless remarked that although "I find good uses for it in the little, I want to try it also in the great, and if it turns out well I will let you know." The meaning here of *im kleinen* and *im großen* remain slightly ambiguous, but it is possible that these terms refer to the "Lesser Work" and "Greater Work," namely, two methods of effecting metallic transmutation, the latter synonymous with making the Philosophers' Stone.[28]

Anthoni Grill's Swedish Success

In 1659, Grill left Amsterdam in poor financial shape. But his story does not end there. He and his family emigrated from the Netherlands and settled in Sweden. The full details of his activities there remain to be uncovered at present; however, it is clear that in Sweden, a land rich in metallic ores (which is presumably what drew him there) Grill's fortunes improved dramatically. Borrichius, once again, provides a report. In 1663, the Danish savant, then in Paris, learned that Grill:

> stirred up partly by the works of Glauber, and partly by his own industry [*propriâ industriâ*], melts out gold and silver with profit from [a mixture of] litharge, iron and copper scoriae, sand (especially the slate from Angers) [...] and a particular red earth, such

[26] Helvetius, *Vitulus aureus*, 832. Indeed, Andries is known to have died in 1665, consistent with Helvetius's account.

[27] Grill to Moriaen, 13 June 1657, transcribed in Moriaen to Hartlib, 22 Jun 1657; HP 42/2/10A–11B. A Latin translation exists at HP 56/1/64A.

[28] "[I]m kleinen finde Ich guten nuzen dabey, wills im großen auch versuchen und so es woll ausfalt will ichs den H[errn] wissen laßen." (HP 42/2/11A). The meaning is clearer in the Latin version: "In minore opere usum ejus deprehendo satis commodum, neque desinam idem probare in majori, et si quid eo proficiam, faciam Te certiorem." (HP 56/1/64B).

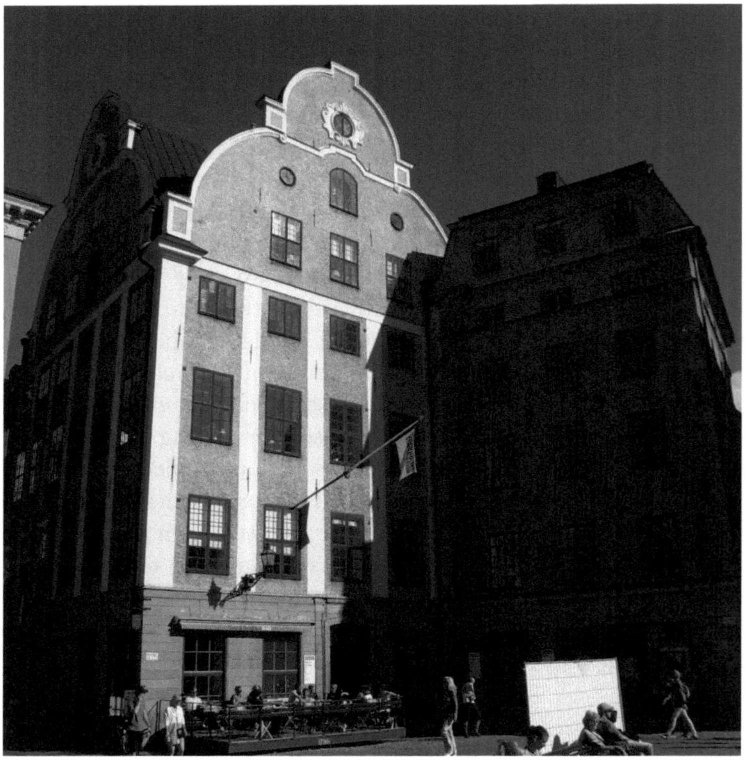

Fig. 4 *Grillska Huset* (the Grill House) in Stockholm, no. 3 Stortorget; the adjoining house on the right, no. 5, was also purchased by Grill in the seventeenth century (Photograph by the author)

> that he obtains more than 40 crowns [*aurei*] a day after having deducted the expenses for sixteen servants and 900 pounds of litharge.[29]

This process is possibly a continuation or development of the work with *potloet* that Grill had attempted while still in Amsterdam. What is clear however is that Grill's social and financial situation must have become quite solid by this time in order for him to run a workshop that employed 16 assistants and that could process 900 pounds of litharge in a day. After Borrichius summarizes Grill's procedure, he records the opinion of "an extremely knowledgeable Frenchman" that this process was not the mere extraction of preexisting precious metals, but a true transmutation of the lead. Applying Helmontian (or perhaps Suchtenian) theory, the unnamed French chymist explained Grill's results as "an exaltation of the lead through the fermentative odor of the admixed scoriae."[30] He added that if Grill had recognized

[29] Borrichius, *Itinerarium*, vol. III, 105–7, 6 October 1663.

[30] On Helmontian theory, see Newman & Principe, *Alchemy Tried in the Fire*, 56–91. Interestingly, this idea of using the "fermentative odor" of copper scoria to effect transmutation was a

this operation as fermentative, and left the mixture in a hotter fire longer, he could have increased his daily profit to 300 crowns, and suggested that Grill should build a bigger furnace outside of some city to effect it. The Frenchman also compared his own experiments with litharge and their results to those carried out by Grill.[31] That Borrichius and a learned French chymist heard about and discussed this matter seriously in Paris indicates that Grill and his chymical work in Sweden had attained a substantial level of international notoriety.

By 1664, Grill had risen to become *Riksguardien vid Kronans Myntverk* (Master of the Royal Mint) in Stockholm. How exactly he attained this prominent state position less than five years after fleeing his creditors in Amsterdam remains to be told, but certainly it required the expertise he had accumulated over the years in assaying, metalworking, and chymical processes involving metals and minerals. Interestingly, when he received the position as mintmaster, he also obtained rights to make and sell gold- and silverware and a license to engage in foreign trade, showing that he had by no means turned his back on his traditional artisanal craft of silversmithing.[32] It is also reported that Grill was involved in lead and silver production at the mines in Stora Skedvi in the province of Dalarna, and tried to extract silver and gold from copper ore.

In 1664, Borrichius was part of a conversation at Henri Justel's (1620–1693) house in Paris where he heard from Abraham Cronström (1640–1696) that "Grill tried new arts in Sweden in vain."[33] Cronström was son of the mintmaster and prefect of the Avesta copper mines in Sweden, and would himself become a mintmaster in Stockholm in 1674. Regardless of Crönstrom's less-than-positive assertion about Grill's unspecified "new arts," it is clear that Grill had achieved considerable success and renown in Sweden, such that he came to the notice (and perhaps the envy) of the large and powerful Cronström family. Indeed, Grill amassed a fortune large enough to purchase a prominent residence on the oldest and most fashionable square in Stockholm, not far from the Royal Mint and the Royal Palace. This house remained in the Grill family for more than two centuries; it still stands at No. 3 Stortorget, and continues to be known today as the *Grillska Huset* (Grill House, Fig. 4). Upon Grill's death in late 1675, Anthoni passed his financial and social success on to his descendants. His son Anthoni (1640–1702), who had come with the family from Amsterdam, followed his father's profession, establishing himself as a goldsmith, jeweller, and merchant in Stockholm, and took up his father's post as mintmaster, as well as becoming engaged in ironworks at Söderfors.[34] His eldest son, also named Anthoni Grill (1664–1727), returned to

pathway investigated by George Starkey—who was also a member of the correspondence network explored in this paper—during the 1650s, Newman & Principe, *Alchemy Tried in the Fire*, 128–35, esp. 130–1.

[31] Borrichius, *Itinerarium*, vol. III, 106–7.

[32] Anrep, *Svenska Slägtboken*, vol. I, 94–5. Borrichius, *Itinerarium*, vol. IV, 99, 27 August 1664 refers to Grill as *monetarius Belga* (the Dutch mintmaster).

[33] Borrichius, *Itinerarium*, vol. IV, 99.

[34] Anrep, *Svenska Slägtboken*, vol. I, 95.

Amsterdam in 1686 where he amassed a considerable fortune, bought a substantial house on the Keisersgracht, became official assayer to the exchange bank in 1712 (presumably having been trained in the practice by his father and/or grandfather), and in 1721 established the *hofje*—an almshouse for the elderly—still known today as Grill's Hofje.[35] Two other sons of Anthoni—Abraham and Carlos—remained in Stockholm where they established and ran one of largest and most successful merchant houses of eighteenth-century Sweden.[36]

After Grill's remove to Sweden, the house and magnificent laboratories he had left behind on the Looiersgracht were rented out in 1660 to none other than Glauber, who lived and worked there until the complex was ordered to be sold in 1661 to pay off Grill's Dutch creditors.[37] It was very probably in these lodgings that the physician, traveller, and scholar Samuel Sorbière (1615–1670) visited Glauber in 1660 and marveled at his multiple ("no fewer than four, at the back of a large house") laboratories: "His laboratories are magnificent, and occupy a wing of his lodgings and the rear of his garden. They are of a prodigious size, and of a completely unique structure."[38] Most or all of what so impressed Sorbière was in fact probably designed and erected by Grill, the construction of which led ultimately to his bankruptcy and emigration to Sweden.

Anthoni Grill thus emerges as an artisan who developed a substantial knowledge—practical and theoretical—of chymistry, both transmutational and more broadly metallic, and thereby elevated himself to a high social and financial situation. Not only was Anthoni capable of designing, carrying out, and improving practical productive processes, but he also operated at a higher epistemic level, reading the books of at least Paracelsus (1493–1541) and Glauber and designing therefrom new experiments and processes *propriâ meditatione* and *propriâ industriâ*. From his initial profession of silver- and goldsmithing and the associated craft of assaying, Grill kept expanding his horizons and expertise to engage in sophisticated and sometimes collaborative chymical projects that reached all the way to that *summum bonum*, the preparation of the Philosophers' Stone, and he designed and built substantial and costly laboratories for his processes and explorations. The knowledge embodied in these workspaces and their instruments were undoubtedly transmitted to his coworkers and laborants, and certainly to Glauber who chose to rent Grill's laboratories almost as soon as they became available. Information and new findings from Grill in Amsterdam were shared with his collaborators, and transmitted by Moriaen into England where they were disseminated through the Hartlib Circle, which at this time included far better-known and widely-published chymists like George Starkey (1628–1665) and Boyle. Likewise, news and details of Grill's Swedish work were discussed as far away as Paris, where

[35] Eeghens, "Grill's Hofje," 51–6; and Kroes, "Nederlands-Zweedse familie Grill," 75–101.

[36] Müller, *Merchant Houses of Stockholm*.

[37] Eeghens, "Grill's Hofje," 50.

[38] Sorbière, *Diverses matieres curieuses*, 176 and 180, in a letter dated 13 July 1660 to Guillaume de Bautru.

one learned chymist found it sufficiently interesting to apply the latest chymical theories to it. The accounts tendered by Borrichius and Helvetius indicate that Andries Grill too was tied into a network of chymists in and outside of The Hague, just as his brother Anthoni was with respect to Amsterdam. Helvetius and others gathered at Andries' workshop to see a remarkable experiment about which he had informed them, and Helvetius used the silversmith's results in his own publication. Additionally, just as Anthoni had been reading and interpreting at least Glauber and Paracelsus, so too his brother Andries read Basilius Valentinus, and connected what he had read to the results of his practical work. Thus it is clear that the two silversmiths were not operating in isolation from other practitioners, nor separated from the textual and other intellectual traditions of chymistry, but clearly exchanged information with these other groups.

Anthoni Grill was almost certainly unusual among silver- and goldsmiths in terms of his chymical sophistication and investigations, not to mention the eventual success he acquired thereby for himself and for his family. One should not, therefore, extend his example too readily to metalworking artisans as a whole; we will need more case-studies to assess accurately the place of such artisans in the chymical world of early modern Europe. Nevertheless, Anthoni and Andries Grill do provide clear and well-documented examples of the chymical activities and the intellectual contacts and contexts that some seventeenth-century artisans could have. Now I will turn to examples provided by goldsmiths elsewhere in Europe.

Parisian Goldsmiths and Chymists

Moriaen and the chymical group around him in the Netherlands that included Anthoni Grill maintained connections to circles wider than the one in England. The recent discovery of a large cache of formerly unidentified manuscripts belonging to the diplomat, adventurer, and natural philosopher Sir Kenelm Digby (1603–1665) reveals that Moriaen was also in direct contact with one of the most important chymists in France at the very same time he was in contact with Worsley, Hartlib, and Starkey in England.[39] One volume of these Digby manuscripts consists of lengthy excerpts copied from notebooks kept by Samuel Cottereau Duclos (1598–1685).[40] Duclos was the first chymist chosen for the Académie Royale des Sciences upon its formation in 1666, and he became a leading figure in that body. He published on chymical composition and the analysis of mineral waters, wrote a substantial treatise on salts, sought the Philosophers' Stone and the alkahest, and presented a lengthy critique of Boyle's ideas on chymistry particularly his

[39] Principe, "Sir Kenelm Digby."

[40] Bibliothèque Nationale et Universitaire de Strasbourg (hereinafter BNU) MS 370.

application of the mechanical philosophy to chymical processes.[41] The Digby manuscript preserves partial transcripts of letters from Moriaen and Glauber to Duclos dating to 1651–1652. Moriaen sent Duclos information about Glauber's processes—he sent some of the very same material simultaneously to Starkey in London—and acted as a conduit for queries from Duclos to Glauber, and Glauber's responses back to Duclos, thus brokering an exchange between the technological entrepreneur Glauber and one of France's most eminent chymists.[42]

The fact that Digby was able to make copies out of Duclos' notebook attests to a close relationship between the two, a relationship corroborated by their collaborative work outlined in other manuscripts of the Digby cache. But these same manuscripts also reveal a wider circle of closely-knit chymists active in Paris throughout the 1650s and 1660s. This circle included, besides Duclos and Digby, the abbé Jullien de Loberie (chaplain of the Collège Fortet) and an abbé Boucaud, Thomas Gobelin (member of a prominent family and counselor to the Paris Parlement), the chymical authors Jean-Baptiste de la Noue, Pierre Borel, and François de Soucy, Sieur de Gerzan, as well as a host of others. The group shared manuscripts, processes, and other information gathered from books or private correspondence, and collaborated (in various combinations) on practical chymical processes, most frequently of a transmutational nature. Some of them frequented well-known intellectual groupings such as the academy of the abbé Bourdelot.[43] The travelling savants Borrichius and Henry Oldenburg (1619–1677) also interacted with the group during their visits to Paris.

Several documents record the participation of goldsmiths in this Parisian chymical circle. One notable example is an experiment recorded initially by Duclos in the early 1650s under the heading *aurum redivivum* and transcribed and annotated later by Digby. The process involves sealing gold hermetically in a flask and exposing it to sunlight and moonlight at particular seasons and then burying it in the ground for extended periods, reportedly causing the gold to undergo various changes to its properties.[44] The annotations in Digby's handwriting clearly indicate that Duclos thought highly enough of the report that he attempted to repeat the experiment himself. What is probably the same process was described some years later by the abbé Boucaud, a close chymical associate of Duclos, who attributed the work to *quidam aurifaber* (a certain goldsmith).[45] Digby annotated his transcription of the process with a cross-reference reading "See among my loose papers, the same processe communicated to me by Monsieur Caillard, who wrought it with the

[41] On Duclos, see Sturdy, *Science and Social Status*, 107–9; Todériciu, "Biographie de Samuel Duclos," 64–7; Stroup, *Company of Scientists*, and "Affaire Duclos."

[42] Principe, "Sir Kenelm Digby," 16–8.

[43] On the Bourdelot Academy, see Brown, *Scientific Organizations*, 231–53.

[44] BNU MS 370, fol. 39v–41v.

[45] Borrichius, *Itinerarium*, vol. III, 134, 7 Dec. 1663, items 4 and 12.

author."[46] This Caillard was himself a goldsmith, identified as such by Boucaud when he describes what is certainly a further development of the same avenue of research—involving burying a hermetically-sealed flask of gold four feet deep on the hill of Montmartre—that was carried out by "Caillard aurifaber."[47] Four generations of goldsmiths named Caillard (or Caillart) are known to have been active in Paris during the sixteenth and seventeenth centuries. The Caillart known to Digby and Boucaud may have been the most well-known of these, Jacques Caillart the Elder, who practiced his trade in Paris from about 1620 until the 1660s. Alternatively, the alchemical Caillart may have been his son, Jacques Caillart the Younger, born about 1622 and active until at least the 1670s.[48]

These accounts are of particular interest for several reasons. First of all, the process for preparing this *aurum redivivum* seems to have been conceived and carried out initially by two Parisian goldsmiths who reported their results to the wider Parisian circle. Repetitions and related experiments with the exposure of gold to sunlight and/or moonlight and subterranean conditions continued within the group—now at the hands of others, often more prominent natural philosophers, for at least a decade after the initial report. Indeed, reports of the first results attracted the attention of several better-known members of the Parisian circle—Duclos, Digby, and Boucaud at least—who proposed modifications to it based on their own practical and theoretical knowledge; these suggestions may well have been transmitted back to the goldsmiths. For example, the initial insistence that particular astrological conditions were necessary seems to have been debunked and discarded. Duclos, for his part, saw the experiment as a route to the Philosophers' Stone that he was seeking: "I believe that this gold thus prepared, being dissolved in the *eau hyliale* (hyleal water) extracted from antimony and mercury sublimate, one part upon six, and cooked according to art and the doctrine of the Cosmopolite, would be rendered highly exalted for the medicine of bodies both human and metallic." (The phrase "medicine of bodies both human and metallic" is a circumlocution for the Stone, which was reputed to be both a universal medicine for human health and a means of "curing" the imperfections of the base metals, thus turning them into gold.) Duclos also suggested that a concave mirror—a device that he and other French chymists were experimenting with at the time—might be used to reflect the rays of sun and moon more powerfully, and questioned whether burial in the earth was really necessary. He even linked the results to the cryptic utterances

[46] BNU MS 370, fol. 39v; Digby's 'loose papers' have not survived, unless they are bound somewhere amid the 6000 pages of the newly-discovered Digby manuscripts.

[47] Borrichius, *Itinerarium*, vol. IV, 267, 25 February 1665.

[48] Bimbenet-Privat, *Orfèvres*, vol. I, 269–70. Jacques Caillard the Elder is mentioned by Michel de Marolles, abbé Villeloin in his contemporaneous recounting of seventeenth-century Parisian goldsmiths, see Marolles, *Livre des peintres et graveurs*, 57 and 126; he is also mentioned in a Parisian baptismal register dating from 1618, *Archive de l'art français*, 324. Some of his designs are depicted in Caillard, *Livre de toutes sortes*.

of the ancient Emerald Tablet of Hermes Trismegestus, writing tersely at the end of his musings, "Pater sol. Mater Luna. nutrix terra aer Vehiculum" ("The father is the Sun, the mother the Moon, the Earth is the Nurse, the air the vehicle").[49] This prominent chymist thus linked the goldsmiths' practice with the theories of Michael Sendivogius ("the Cosmopolite") and Hermes. Duclos's ideas may well have cycled back into modified trials by others in the circle, and very possibly by the goldsmiths as well.

Further insight on the place of goldsmiths in the Parisian chymical scene is provided by diary entries made by Borrichius from 1663 to 1665. Amid his frequent visits to natural philosophers, professors, theologians, and various well-known Parisian *curiosi*, Borrichius also made no fewer than eight visits to an "ingenious goldsmith and jeweller" and "noteworthy investigator" [*curiosus insignis*] by the name of Rosselle, who also visited him twice in return.[50] This "Rosselle" is presumably one of five gold- and silversmiths by the name of Roussel known to have been active in Paris in the 1660s. Borrichius' description of him as "goldsmith and jeweler" suggests that he was Claude Roussel (d.1678) who bore the title of *orfèvre joaillier ordinaire du roi*, although this identification must remain provisional.[51] On some visits Roussel displayed marvels of artifice, on others he showed devices (such as a special lamp furnace) or techniques (such as a simple method of assaying minerals) of his own design. Most frequently, the topic of conversation was chymistry, sometimes pharmaceutical in nature but more usually transmutational. Some of these processes Roussel had devised himself, some he had gathered from others. The information he relayed clearly attests to the goldsmith's connections and exchanges with a range of contacts both in Paris and across Northern Europe. He tells for example of having purchased four ounces of chymically-produced gold from an unspecified German court, and of having received eight grains of chymical gold from the *dispensator domus* of the Duchesse d'Aiguillon (1604–1675).[52] If this last report is accurate, these few grains of gold would certainly have been scraped from a fragment of gold supposedly made by the (in)famous Noël Picard, called Dubois, in 1636 at the court of Louis XIII. The Duchesse had obtained the gold from her uncle, Cardinal Richelieu, who was present at the transmutation, and who eventually had Dubois imprisoned and hanged when he could not make good on his promise to provide the cash-strapped monarch with a weekly supply of precious metal.[53]

---

[49] BNU MS 370, fol. 41r–v.

[50] "Invisi Dn. Rosselle in plateâ Bussi, curiosum aurifabrum et gemmarium [...]." (Borrichius, *Itinerarium*, vol. IV, 60); on 107 he is called "curiosus insignis"; further visits or information occurs on 131–3, 143–4, 165, 168, 176–7, 220, 275, 285.

[51] On the various Roussels, see Bimbenet-Privat, *Orfèvres*, vol. I, 500.

[52] Borrichius, *Itinerarium*, vol. IV, 108, 6 September 1664. Roussel's lamp furnace is mentioned and sketched on 132–3, 20 September 1664.

[53] On Dubois, see Principe, "Sir Kenelm Digby," 11–4; note there that the sentence tagged with footnote 29 should read 'Olaus Borrichius' not Pierre Borel.

On one occasion, Roussel reported that another goldsmith (located on the Place Dauphine) told him that someone in Paris had recently transmuted 40 ounces of silver into gold, and that the goldsmith possessed a piece of this gold.[54] Roussel also relayed information to Borrichius from Duclos (regarding the latter's production of red glass), confirming that the goldsmith was indeed linked into the Parisian circle of chymists. Likewise, on 22 November 1664, Borrichius found Roussel in the company of "a certain Flemish chymist named Regius" and the three discussed Van Helmont (1579–1644) together.

Goldsmiths and Chrysopoeia in England

Several connections complete the third side of the Amsterdam-Paris-London triangle of chymical exchange. Digby of course maintained correspondence with his countrymen at home, and, more significantly, travelled back to England in 1654–1655 when it seemed briefly possible that Catholics would be tolerated by the Commonwealth. During this time, Digby continued his chymical interests in direct contact with Hartlib and members of his circle such as Clodius, George Starkey (1628–1665), and Robert Boyle (1627–1691), to whom he communicated results and observations gathered in Paris. At present I have not uncovered any goldsmiths who contributed actively to the chymical conversations of Hartlibians in London in ways akin to what Grill did in the Netherlands, and Caillart, Roussel, and others did in Paris. Nevertheless, interactions of gold- and silversmiths with members of the London circle are not entirely absent, and serve to illustrate an alternate way in which goldsmiths were involved in the chymical world.

Probably the most talented chymist of the English branch of this international network was Starkey, the Bermuda-born and Harvard-educated emigré who had arrived in London from Massachusetts in 1650 and quickly became a celebrity among Hartlibians and a particular friend and teacher in chymical matters to Boyle. Starkey was a microcosm of the broad chymical world of the seventeenth century. He was talented in both theory and practice, and applied himself to virtually every branch of chymistry: he sought the Philosophers' Stone and metallic transmutation, he endeavored to prepare new chymical medicines, he made ends meet (sometimes) by manufacturing perfumes, cosmetics, essential oils, and distilled liquors, by practicing medicine, and by working in mining and refining operations.[55] It was his potentially lucrative transmutational chymistry—and his supposed relationship with a successful adept named Eirenaeus Philalethes—that particularly attracted attention, and it is in this context that gold- and silversmiths appear.

---

[54] Borrichius, *Itinerarium*, vol. IV, 132.

[55] Newman, *Gehennical Fire*; Newman & Principe, *Alchemy Tried in the Fire*; and Starkey, *Alchemical Laboratory Notebooks*.

In 1651, Hartlib reported that Starkey had successfully prepared silver from antimony. But Starkey's silver "was in weight equal to Gold," that is, it was the chymical product called *luna fixa*, a metal with the appearance of silver, but the density and chemical properties of gold. The three collaborators—Worsley in London and Moriaen and Grill in Amsterdam—Hartlib reported, "undertake to turne that Antimonial silver into Gold."[56] One of Digby's associates, Dr. Farrar, soon appeared, or in Starkey's own words, "Came Gaping" to Starkey's laboratory to see for himself, and reportedly attempted (unsuccessfully) to purchase the process for the princely sum of 5,000 pounds. Yet despite such a promising process, from which Hartlib and his associate John Dury reckoned Starkey could "easily make [...] 300 lb a year," Starkey still encountered difficulty. As he reported to Boyle, the weight of the silver

> did distract the Cockscombes of almost 20 refiners & Goldsmiths to whom I showed the silver, who all liked it, but stood amazed at the weight, & so desired an assay of it, at the publique hal, which I was unwilling to, because of publiquenes, for every one (not one in at least 16 missing) did say it was Philosophick silver, insomuch that I Could have wished it in the Sea.[57]

Hartlib recorded that Starkey

> was greatly vexed to have to abandon a lump of his silver at the goldsmith's. Several of them marveled that it was so unlike other silver. Finally he met one who paid him what he asked. But Mr. Worsley advised that it was dangerous to sell it to a goldsmith and that it should be brought to the mint.[58]

This account even reappears, with some embellishment, in *Secrets Reveal'd*, a cryptic, extremely popular, and posthumously published work about making the Philosophers' Stone that Starkey wrote under the persona of his fictive adept Eirenaeus Philalethes. There "Philalethes" laments the fate of the successful transmuter, for he will find it dangerous or impossible to sell any gold or silver he makes because the goldsmiths will always detect its superiority to ordinary gold or silver and demand to know where it came from. Philalethes recounts that once when he tried to sell some silver that he had transmuted, the goldsmith immediately said it was chymical; "which when we heard, we prively withdrew, and left both the Silver and the price of it, never more demandable."[59] Thus in Starkey's case, the expertise of gold- and silversmiths was seen not as a source of information or collaboration, but rather as an obstacle to making a comfortable alchemical living.

---

[56] Hartlib, *Ephemerides* 1651, HP 28/2/18A.

[57] Starkey to Boyle, c. April/May 1651, Starkey, *Alchemical Laboratory Notebooks*, 20.

[58] Hartlib, *Ephemerides* 1651, HP 28/2/18B. Curiously this is virtually the only entry in the *Ephemerides* that Hartlib wrote in German, perhaps to preserve secrecy about such an apparently dangerous undertaking. When in 1653 Starkey again succeeded in having a goldsmith buy his antimonial silver, he obtained the substantial price of 40 shillings an ounce for it, almost eight times the usual rate for ordinary silver; Hartlib, *Ephemerides* 1653, HP 28/2/68A.

[59] Starkey, *Secrets Reveal'd*, 39.

Conclusion

The international network of alchemical correspondents and collaborators outlined in this paper contributes to our understanding of seventeenth-century alchemy's content and context in several ways. First, it is clear that several gold- and silversmiths were active members of the network. Rather than merely being a source of random observations made casually in the course of exercising their craft, they initiated projects, conducted experiments, made innovations, read learned treatises, and communicated or collaborated with more traditionally-educated or more natural philosophically inclined members. While someone like Anthoni Grill was undoubtedly an exceptional figure, he nevertheless serves to illustrate one end of the spectrum of possible social and intellectual positions occupied by early modern gold- and silversmiths. His active participation and eventual notoriety and success indicate the social boundaries that were crossed within the network, a situation that also existed with Caillart and Roussel and others in the French context as well. These examples thus begin to resituate the role of such artisan-chymists within the larger ambit of chymistry. While not erased, the neat division of artisan and natural philosopher emerges considerably blurred.

The diversity of characters within the network and their various connections and collaborations revises still-prevalent views of alchemical laborers as characteristically solitary, secretive, and isolated. Indeed, when we consider the reports of Grill's multiple laboratories staffed with at least 16 operators, it is hard not to recall the contemporaneous depictions of chymical laboratories in Netherlandish genre painting that almost always depict a foregrounded main experimenter with a group of other workers in the background tending various processes. While there are certainly purely stylistic and iconographic motivations behind such depictions (similar layouts are used in other genre paintings not showing chymists), the accounts of Grill's workshop do add a level of verisimilitude to these artistic representations that we might not otherwise have attributed to them. The image of a busy, and presumably compartmentalized and hierarchical, chymical workshop pulls our understanding of these locales towards a comparison with contemporaneous artisanal ateliers, which, if the principals involved were themselves artisans, would have been the obvious model upon which to draw virtually automatically. (Interestingly, a modern academic chemical research group retains much of this character and structure.) The transconfessional and international character of the larger network—which included Catholics and assorted types of Protestants, as well as Dutch, French, English, Danes, Swedes, and Germans—parallels the way in which it cut across social levels. This situation complicates some of the standard ways in which historians are accustomed to draw reckoning lines, and recalls the ideal of a *respublica litterarum* even in a subject as distant from traditional humanism as sooty chymistry.

Bibliography

1855–1856. *Archive de l'art français*, 6 vols, ed. Anatole de Montaiglon. Paris: Dumoulin.

Anrep, Gabriel. 1871–1875. *Svenska Slägtboken*, 3 vols. Stockholm: Haeggström.

Bimbenet-Privat, Michéle. 2002. *Les Orfèvres et l'orfèvrerie de Paris au XVIIe siècle*, 2 vols. Paris: Commission des Travaux Historiques de la Ville de Paris.

Borrichius, Olaus. 1983. *Itinerarium 1660–1665*, 4 vols. Copenhagen: Danish Society of Language and Literature.

Brinkman, A.A.A.M. 1982. *De Alchemist in de prentkunst*. Amsterdam: Rodopi.

Brown, Harcourt. 1934. *Scientific Organizations in Seventeenth Century France*. Baltimore: Williams & Wilkins.

Caillard, Jacques. 1627. *Livre de toutes sortes de feuilles pour servir à l'art d' orfevrie de l'invention de Jacques Caillard*. Paris.

Eeghen, I.H van. 1970. Het Grill's Hofje. *Jaarboek van het Genootschap Amstelodamum* 62: 49–86.

Gelder, H.E van. 1937. Een Haagsche bekerschroef van Andries Grill 1642. *Mededeelingen van den Dienst voor Kunsten en Wetenschappen der Gemeente 's–Gravenhage* 4: 113–117.

Hartlib, Samuel. 2002. *Hartlib Papers*. Sheffield: HROnline, Humanities Research Institute, University of Sheffield.

Helvetius, Johann Friedrich. 1678. *Vitulus aureus*. In *Musaeum hermeticum reformatum et amplificatum*, 815–863. Frankfurt.

Kroes, Jochem. 2012. De Nederlands-Zweedse familie Grill en haar Chinese wapenserviezen. *Jaarboek van het Centraal Bureau voor Genealogie* 66: 75–101.

Leng, Thomas. 2008. *Benjamin Worsley (1618–1677): Trade, Interest and the Spirit in Revolutionary England*. Woodbridge: Boydell Press.

Lennep, Jacques van. 1966. *Art et alchimie. Étude de l'iconographie hermétique et de ses influences*. Brussels: Éditions Meddens.

Lorm, Jan Rudolph de. 1999. *Amsterdams goud en zilver*. Amsterdam: Waanders Uitgevers.

Marolles, Michel de. 1872. *Le livre des peintres et graveurs*. Paris: Daffis.

Moran, Bruce T. 2005. *Distilling Knowledge: Alchemy, Chemistry, and the Scientific Revolution*. Cambridge, MA: Harvard University Press.

Müller, Leos. 1998. *The Merchant Houses of Stockholm c.1640–1800: A Comparative Study of Early-Modern Entrepreneurial Behaviour*. Uppsala: Uppsala University Library.

Newman, William R. 1994. *Gehennical Fire: The Lives of George Starkey an American Alchemist in the Scientific Revolution*. Cambridge, MA: Harvard University Press.

Newman, William R., and Lawrence M. Principe. 1998. Alchemy vs. Chemistry: The Etymological Origins of a Historiographic Mistake. *Early Science and Medicine* 3: 32–65.

Newman, William R., and Lawrence M. Principe. 2002. *Alchemy Tried in the Fire: Starkey, Boyle, and the Fate of Helmontian Chymistry*. Chicago: University of Chicago Press.

Nummedal, Tara E. 2007. *Alchemy and Authority in the Holy Roman Empire*. Chicago: University of Chicago Press.

Pijzel-Dommisse, Jet. 2005. *Haags goud en zilver: Edelsmeedkunst uit de Hofstad*. Zwolle: Waanders Uitgevers.

Principe, Lawrence M. 1998a. Diversity in Alchemy: The Case of Gaston 'Claveus' DuClo, a Scholastic Mercurialist Chrysopoeian. In *Reading the Book of Nature: The Other Side of the Scientific Revolution*, eds. Allen G. Debus and Michael Walton, 181–200. Kirksville: Sixteenth Century Journal Publishers.

Principe, Lawrence M. 1998b. *The Aspiring Adept: Robert Boyle and His Alchemical Quest*. Princeton: Princeton University Press.

Principe, Lawrence M. 2000. The Alchemies of Robert Boyle and Isaac Newton: Alternate Approaches and Divergent Deployments. In *Rethinking the Scientific Revolution*, ed. Margaret J. Osler, 201–220. Cambridge: Cambridge University Press.

Principe, Lawrence M. 2013a. *The Secrets of Alchemy*. Chicago: University of Chicago Press.

Principe, Lawrence M. 2013b. Sir Kenelm Digby and His Alchemical Circle in 1650s Paris: Newly Discovered Manuscripts. *Ambix* 60: 3–24.

Principe, Lawrence M., and Lloyd DeWitt. 2002. *Transmutations: Alchemy in Art*. Philadelphia: Chemical Heritage Foundation.

Smith, Pamela H. 2000. Vital Spirits: Redemption, Artisanship, and the New Philosophy in Early Modern Europe. In *Rethinking the Scientific Revolution*, ed. Margaret J. Osler, 119–135. Cambridge: Cambridge University Press.

Smith, Pamela H. 2004. *The Body of the Artisan: Art and Experience in the Scientific Revolution*. Chicago: University of Chicago Press.

Sorbière, Samuel J. 1660. *Relations, lettres, et discours sur diverses matieres curieuses*. Paris.

Starkey, George [Eirenaeus Philalethes]. 1669. *Secrets Reveal'd, or an Open Entrance to the Shut-Palace of the King*. London.

Starkey, George. 2004. *The Alchemical Laboratory Notebooks and Correspondence*, eds. and trans. William R. Newman and Lawrence M. Principe. Chicago: University of Chicago Press.

Stroup, Alice. 1990. *A Company of Scientists: Botany, Patronage, and Community at the Seventeenth-Century Parisian Royal Academy of Sciences*. Berkeley: University of California Press.

Stroup, Alice. 2002. Censure ou querelles scientifiques: L'affaire Duclos (1675–1685). In *Règlement, usages et science dans la France de l'absolutisme*, eds. Christiane Demeulenaere-Douyère and Éric Brian, 435–452. Paris: Tec et Doc Lavoisier.

Sturdy, David. 1995. *Science and Social Status: The Members of the Académie Royale des Sciences, 1666–1750*. New York: Boydell Press.

Todériciu, Doru. 1974. Sur la vraie biographie de Samuel Duclos (Du Clos) Cotreau. *Revue d'histoire des sciences* 27: 64–67.

Turnbull, George Henry. 1947. *Hartlib, Dury and Comenius: Gleanings from Hartlib's Papers*. London: University Press of Liverpool.

Webster, Charles. 1976. *The Great Instauration: Science, Medicine and Reform, 1626–1660*. New York: Holmes & Maier Publishers.

Young, John T. 1998. *Faith, Medical Alchemy and Natural Philosophy: Johann Moriaen, Reformed Intelligencer, and the Hartlib Circle*. Brookfield: Ashgate.

Igne refutata: Thermal Analysis in the Laboratory Practices of John Dwight and Ehrenfried Walther von Tschirnhaus

Morgan Wesley

Abstract The creation of European porcelain was long speculated to have originated from the flow of information from Chinese and Japanese sources; however, no substantive evidence of direct knowledge transfer has been discovered. This paper endeavors to shift the focus from an externally driven developmental process and relocate the principal method of innovation within the experimental framework established by early modern chymistry. Evidence for the use of thermal experimentation will be considered as a foundational element toward a chymical solution to the problem of porcelain production. Excavated material from the workshop of the seventeenth century English arcanist John Dwight and the published experiments of the seventeenth century Silesian natural philosopher Ehrenfried Walther von Tschirnhaus will provide the basis for this examination. This material, along with the thermal elements unique to the successful Meissen porcelains will be used to frame the initial comparisons of the technological differences between the European and Far Eastern productions, and serve to pose further questions regarding the impact of experimental techniques and limitations on European porcelain arcanistry.

Until recently the study of early modern ceramics was principally the domain of art historians and archaeologists, limiting the scope of questions being posed by academics. Increasingly the role of natural philosophy, medieval alchemy and nascent chemistry in ceramic evolution has been revealed, bringing with it the necessity to align discourse on ceramic practice with the discipline of history of science, as art history lacks the framework to pose questions that would advance the discussion. This paper is an exploratory work that builds the foundations to bridge those separate traditions to expedite a more consilient examination of ceramic technology and innovation. The principal focus is on the adoption of specific

M. Wesley (✉)
Department of History, University of Oxford, Oxford, UK
e-mail: thenex@mac.com

S. Dupré (ed.), *Laboratories of Art*, Archimedes 37,
DOI 10.1007/978-3-319-05065-2_8,

chymical practices in early modern ceramic innovation that lead to a drastic reshaping of the experimental landscape. While many adopted practices flowed seamlessly between the two disciplines through shared tools and techniques developed by generations of practitioners working with fire, the specific nature of ceramics forced adaptations on other practices. Elements borrowed from controversial chymical fire analysis found more specific application in investigating thermal qualities of ceramic bodies in laboratory experimentation. Used in ceramic investigations in England and Saxony during the late seventeenth and early eighteenth centuries, this hybrid approach was a defining moment in high-fire European ceramic production.

The recent change in terminology surrounding medieval and early modern distinctions between alchemy, chymistry and chemistry is particularly significant in informing the need for similar clarifications in discussions of historic porcelains.[1] As a more extensive network of documentary sources related to early modern European ceramics is assembled, the mercurial nature of the term 'porcelain' is increasingly cast into the spotlight. From the traditionally ascribed basis in records of Marco Polo's (c.1254–1324) voyage, different interpretations have existed, and no substantial historic reconciliation has been attempted. It is beyond the scope of this paper to attempt to do so, but establishing a basic set of terms is necessary. The term 'porcelain' will be employed to denote white, translucent, high-firing ceramic bodies, consistent with the material described by Polo and exported from China into Europe.[2] Two subdivisions will further be mentioned in this text; those objects made through a combination of kaolinized clay and alkali salt rich stone will be referred to as 'hard-paste', while objects made of a combination of clay and ground glass, sand, or quartz flux will be viewed as 'soft-paste'. These are entirely modern separations, used here for clarity, and specific examples given in the following paper will demonstrate they are not historic terms.

The seventeenth and eighteenth century quest to duplicate Chinese porcelain borrowed not only techniques from the chymical tradition, but even notions of higher *arcana*. This notion of the adept's knowledge dovetails with the terminology employed within the Saxon court of Augustus II the Strong (1670–1733) as progress towards commercial European porcelain was made. The term *arcanum* is employed in court records and correspondence as it related to the porcelain secret, and those individuals working to produce porcelain bodies were certainly possessed of substantial industrial secrets, making them, fittingly, *arcanists*.[3] While scholars have pointed out that the term *arcanum* is not strictly limited to alchemical pursuits, being widely employed within the book of secrets tradition during the later

---

[1] Newman & Principe, "Alchemy vs. Chemistry."

[2] Originally composed as *Livres des merveilles du monde*, the original manuscript was purportedly composed by Rustichello while he shared imprisonment with Marco Polo after the latter's return to Venice in 1295 AD. A full discussion of Polo's references to porcelain can be found in Carswell, *Blue and White*, 52–4.

[3] For the full correspondence, see Tschirnhaus, *Amtliche Schriften*, ch. II.

medieval and early modern periods, in the context of porcelain, strong arguments can be made towards the artisanal-alchemical bridge. The correspondences that remain in the archives at Dresden, originate from that alchemical tradition, to the extent that arguments have been leveled to suggest that the correspondence was, in fact, about the creation of the Philosophers' Stone rather than a search for the composition of porcelain. While later *arcanists* cannot be seen to participate in transmutational alchemy on more than an individual level, the origination of the term in the late seventeenth and early eighteenth centuries is clearly linked to those pursuits. Further, it should be noted that these first *arcanists* are entirely separate from communities and guilds of potters in their various regions, offering an entirely novel actor class that appears at the end of the seventeenth century, chymists whose focus was on the production of translucent ceramics rather than metallurgical, medical, or other natural philosophical avenues.

No evidence towards production of a hard-paste porcelain body by European potters before the eighteenth century is extant, leaving the chymist's laboratory as the principle space of ceramic innovation. Specific investigation of the physical properties of Chinese porcelain undertaken by John Dwight (c.1633–1704) and Ehrenfried Walther von Tschirnhaus (1651–1708) represent the first successful examples of chymistry's methodological apparatus being directed toward porcelain innovation. Thermal experiment was an important element in the practices of both individuals and evidence presented in the archaeological record demonstrates Dwight's re-firing of Chinese sherds while Tschirnhaus published results of the experimental application of burning lenses to determine the physical properties of varied ceramic bodies. Dwight was ultimately unsuccessful in his endeavors, likely due to economic realities, while Tschirnhaus's work would lay the cornerstones for the eventual production of Meissen porcelain.

The Potter's Art as Craft Practice

To understand why the shift from the intellectual space of the craft workshop to the chymical laboratory was critical to the advancement of European ceramic development, an examination of the awkward position of pottery within the hierarchy of the decorative arts before the early modern period is worthwhile. Prior to its interaction with chymistry, the relationship between the potter's craft and fire was principally of a mechanical, fixative nature. As the heat of the kiln was known to remove the water from earthen bodies, its ability to affect permanent transformation on the materials placed in the kiln was its principal contribution to the success of ceramic production. Those objects existed first as trade goods rather than artisanal objects and only later became valued as aesthetic products. Vannoccio Biringuccio's (1480–c.1539) introduction to his brief "discourse on the art of the potter" from his *Pirotechnia* of 1540, encapsulates both the recent recognition received by ceramic craft and its inseparability from fire:

> Having started to tell you of working potter's clay for making crucibles and shells, the wish came to me to tell you of the practice of this art also. Although it may seem at first glance to be outside the order and purpose of my writing, he who considers well will see that it is not unrelated to it but proper, since it is wholly dependent on the agency and power of fire if it is to be brought to its perfection. Moreover, the potter's glazes and colors are all substances of various metals or impure minerals and therefore belong to fire. Since it is my intention to treat of fire, minerals, and metals for you, I surely should not have omitted this, particularly because it is a necessary art which enriches and is greatly praised both for its ingenuity and its beauty.[4]

Like the other arts treated in the *Pirotechnia*, the action of the fire on base earths and minerals provided integral transformation in the objects shaped by the potter's hands. At the time of Biringuccio's writing, the potter's art was on the threshold of a paradigm shift that would create a marked division between craft knowledge and theory based practice, setting two separate courses for ceramic innovation in the later sixteenth and seventeenth centuries. His citation of alchemy as one of "two sources as [pottery's] principal basis," is referential to the provision of purified minerals and "elemental mixtures."[5] However, it also foreshadowed the intellectual engagement of chymists in the quest for the production of European porcelain, bringing the weight of the older alchemical tradition to bear.

Prior to the fourteenth century, European pottery was external to the hierarchy of the esteemed arts, such as metallurgy, glassmaking, painting, and dyeing. The extensively copied treatise *De diversis artibus*, composed in the thirteenth century by the Benedictine monk Theophilus, thoroughly introduces the full range of artisanal productions valued in Europe in its first book.[6] Pottery is entirely absent from this discussion of laudable arts and only appears once in the text as a canvas for enameling.[7] This is a continuation of pottery's position in the classical period, when earthen objects were valued as trade goods, everyday wares, or a medium for other decoration, not for their individual aesthetic form.[8]

It was not until advancements in the production of luxury quality European pottery during the fifteenth and sixteenth centuries that the cultural perception of the material changed significantly, driving engagement by the upper classes and natural philosophers. An example can be found in a letter from Lorenzo de Medici (1449–1492) to the Malatesta family in 1490, favorably comparing their gift of *maiolica* to silver objects, which is one of the earliest pieces of documentary

---

[4] Biringuccio, *Pirotechnia* (1942), 392.

[5] Biringuccio, *Pirotechnia* (1942), 392.

[6] For various editions of the treatise, see Theophilus, *Essay Upon Various Arts* (1847), *The Various Arts* (1961), and *On Diverse Arts* (1963).

[7] "Quam si diligentius perscruteris, illic inuenies quicquid in diuersorum colorum generibus et mixturis habet Graecia, quicquid in electrorum operositate seu nigelli uarietate nouit Ruscia, quicquid ductili uel fusili seu interrasili opere distinguit Arabia, quicquid in uasorum diuersitate seu gemmarum ossiumue sculptura auro decorat Italia, quicquid in fenestrarum pretiosa uarietate diligit Francia, quicquid in auri, argenti, cupri et ferri lignorum lapidumque subtilitate sollers laudat Germania." (Theophilus, *The Various Arts* (1961), 4).

[8] On the relative value of the potter's work, see Boardman, "Trade in Greek Decorated Pottery."

evidence demonstrating the elevation of pottery into the canon of decorative arts.[9] Within the next two decades, the production of new types of earthenware spread across Italy rapidly replacing vernacular wares, such as those traditionally produced in Orvieto and Montelupo.

As the sixteenth century progressed the demand for porcelain, introduced in small quantities during the fourteenth and fifteenth centuries, exploded alongside a burgeoning maritime trade. While the first successful efforts to produce a ceramic body that replicated some of porcelain's qualities were those of the Medici workshops between 1575 and 1587, they failed to produce hard paste porcelain that was simultaneously durable and heat resistant.[10] Medici porcelain has long been connected to the alchemical interests of Francesco I de Medici (1541–1587) though it required external aid. This outside knowledge was provided by an unknown Levantine leading to the production of a glassy porcelain body that is closely related to Islamic fritwares, but not Chinese material.[11]

It would take another 130 years before the commercialization of a hard paste porcelain body at the Meissen factory in Saxony. The reason for this significant lag in production cannot be connected to a lack of proper resources or technology, as highly kaolinized clay sources, identical to the China clay that the Jesuit priest Père Francois Xavier d'Entrecolles (1664–1741) would write about in 1712, were exploited for the production of Hessian crucibles for laboratories across Europe as early as the fourteenth century.[12] Arguments that the recipes and manufacturing processes of crucibles, such as those produced at Hesse, would have been closely guarded secrets must be considered a contributing factor in the reduced awareness of European potters as to the suitability of such clay for the production of a high firing body. Those arguments can only be validated by assuming that ceramics did indeed exist apart from the other arts of fire at this period; for observations by chymists, metallurgists and glassmakers on the properties of the crucibles, and indeed the entire basis for the valuable export trade of Hessian crucibles, was of their highly refractory, hard firing nature.

The situation again suggests that the lack of a craft practice-based solution to the production of porcelain hindered progress in European investigations, in contrast to the gradual evolution of the Far Eastern wares through extended refinement within pottery communities in China and later Korea and Japan.[13] Potters in those regions, through wider access to materials and support were able to utilize the full range of

[9] For a specific discussion of the social standing of this material, see Wilson, "Le maioliche."

[10] Kingery & Vandiver, *Masterpieces*, 141.

[11] Liverani, *Porcellane dei Medici*, 8, 47, discusses the mysterious Levantine, while referring to Kingery & Vandiver, *Masterpieces*, 141, for related material analyses and comparisons with contemporary examples Islamic fritwares and Chinese porcelains. Further material can be found in Kingery & Vandiver, "Medici Porcelain."

[12] d'Entrecolles, "Lettres." Regarding the use of kaolinized clay for the production of Hessian wares, see Martinón-Torres, Rehren & Freestone, "Mystery of Hessian wares."

[13] For the various avenues of incremental development of Chinese porcelains, see Kerr & Wood, *Ceramic Technology*, 146–63.

clay native to the region, slowly perfecting what was considered an important art form. The long period of development in the Far East allowed incremental refinements necessary for successful innovations through craft practice. With a limited time span driven by the demand for the material in Europe, the interrogative techniques of the chymist's laboratory were necessary to circumvent the slower, less focused workshop process.

Experiment and Fire Analysis

As discussed, the primary hindrance on the production of a porcelain body was the lack of identified materials that could be combined to produce these translucent, high fire, ceramic objects. The production of material at Hesse and the later discovery of kaolinic clay in Cornwall, England, again demonstrate that Europe did not lack the raw materials, but rather that potters were unaware of the properties of those materials, even though other artisans were making use of them. The failure was thus not connected to any form of geological predetermination, but entirely by social constraints limiting the application of knowledge to the problem.

The speculation that porcelain was comprised of either special clay, fermented for generations, or of crushed shell incorporated in its manufacture, was a direct indication that similar materials were outside the realm of contemporary craft knowledge.[14] In order to successfully move past these limitations, the experimental techniques of the natural philosophers allowed objects from China to serve as prototypes for interrogation towards their properties and composition. It is at this point that the techniques employed by chymistry, some borrowed from metallurgical assaying and the furnaces of the glassmakers in the centuries-long discourse between the disciplines, came into full application in the pursuit of ceramic innovation.

Due to the highly resistant nature of porcelain to analysis via distillation, the application of direct heat was the recourse of both Dwight and Tschirnhaus. While their techniques were radically different, applying the heat of the furnace and the energy of the sun respectively, the basic principles of this thermal experiment was closely linked to the long-standing application of chymical fire analysis.

The exposure of both Dwight and Tschirnhaus to fire analysis during their studies at university can be situated within a larger debate regarding the legitimacy of the process during the seventeenth century. The ultimate goal of separating bodies into their constituent components and, ideally, recombining them through the application of fire, was assailed on a methodological and philosophical level in

[14] The speculation regarding the extended fermentation of clay and the addition of 'artificial minerals' to Chinese porcelain comes from Bacon, *New Organon*, Book II, Aph. 50.

the writings of the chymists.[15] Supporters, such as Daniel Sennert (1572–1637), argued that it presented the most effective means of resolution, provided that violent reactions were avoided in favor of distillation and more subtle applications of heat. While Francis Bacon (1561–1626) described the action of fire on the principles of a body as something that "confounds" in that "many natures which are in fact newly brought out and superinduced by fire and heat."[16] Similarly Robert Boyle (1627–1691) questioned the role of fire in analysis, based on the uncertainty of its ability to alter rather than separate substances.[17] This position of Boyle's can be viewed as somewhat problematic, considering the American chymist George Starkey's (1628–1665) extensive influence on his early research. Starkey's notebooks contain extensive application of fire analysis to test various experiments, with the notation *refutata per ignem* appearing multiple times.[18]

Debus clearly argues that the debate over the acceptability of fire analysis was closely linked to the ongoing conflict being waged against Aristotelian thought in the seventeenth century.[19] As such many of the arguments against, from an empirical standpoint, must be viewed as couched within that particular political rhetoric. Within the practices of Dwight and Tschirnhaus, their education in chymistry points to a profound awareness of the controlled application of fire in laboratory practice, but more practical demands circumvented their full engagement within the Aristotelian debate and records of their work support the notion that fire became a critical tool for interrogation of material, but as commentators have pointed out, no true fire analysis could occur.

Considering the context of ceramics, the limiting factor of fire analysis must necessarily be in its inability to undo changes in crystalline structure during the initial firing of an object. Because of these phase shifts, the specific components of a ceramic body resist resolution by thermal or chemical means, making the original materials inscrutable. Even with modern scanning electron microscopy only an analysis of the volumes of base elements is available (though insight into crystalline structure is provided); establishing a definitive source or recipe for the ware prior to firing is frequently impossible. Dwight and Tschirnhaus were subject to substantially greater limitations, with the possible attributes available for their analysis restricted to: temperature of vitrification, point of liquefaction and failure of the body, separation of glaze and body material in phases, and reaction of any applied decorations in a manner different to the body or glaze. This limited access to deeper structures would demand that any speculation as to source or recipe would have to

[15] The essay by Debus, "Fire Analysis," encapsulates much of the instability related to fire analysis as it appears in the major texts of the seventeenth century, while Newman, *Promethean Ambitions*, 251–62, revisits and furthers the discussion as it relates to the art-nature debate.

[16] Bacon, *New Organon*, Book II, Aph. 7.

[17] Boyle, *The Sceptical Chymist.*

[18] For more on Boyle and Starkey's relationship and the work of Starkey as an experimentalist, see Newman & Principe, *Alchemy Tried in the Fire*, 92–128; and Newman, *Gehennical Fire*.

[19] Debus, "Fire Analysis," 128–30.

be made with comparative data from other experiments on local raw materials, a route that both men pursued aggressively, as discussed below.

That Bacon and Boyle were key figures in the debate regarding fire analysis had a corollary impact on the advancement of ceramic innovation, and perhaps on the ultimate failure to discover the *arcanum* in England. As the evidence of Dwight's experimentation is considered, his links to Boyle and the influence of Bacon's writings become increasingly important. Due to the restrictions of his fire analysis, he was unable to disprove Bacon's speculation that there was an 'artificial mineral' employed in the production of porcelain, and his later experiments and workbooks demonstrate that while he became intimately aware of the firing temperatures connected to the Chinese prototypes he was able to secure, he never gave up on the idea of a synthetic additive that would make clay into porcelain.

John Dwight's Fulham Pottery: Chymical Beginnings

Dwight was the first individual to record limited success at producing porcelain in England. Born sometime between 1633 and 1635 in Gloucestershire, he was able to secure a place at Oxford around 1655, leaving by 1660/1.[20] Although officially reading for a Bachelor's of Ecclesiastical Law during this period, Dwight was employed in the laboratory of Boyle.[21] The influence of this period crucially shaped the course of the young Dwight's life, as seen in his admission as an old man to Sir John Lowther, F.R.S. (1642–1706), who recounted in a letter to William Gilpen, dated 12 March 1697/8: "[Dwight] gives this acct of himself yt he was bred at ye University studyed Civil Law & Physick a little, but most Chymistry [. . .]."[22]

This period in Boyle's laboratory corresponds with Robert Hooke's (1635–1703) employment there as an assistant and Dwight and Hooke remained friends throughout their lives. It also corresponds with the period that saw the influence of Johann Rudolph Glauber's (c.1604–1670) *Furni novi philosophici* on Boyle and the employment of various analytical techniques shared between Boyle and Starkey.[23] Dwight was fluent in both Latin and English and would have had access to the full variety of texts moving through the laboratory. It is reasonable to speculate that, as this period coincided with Boyle's increasing blindness, Dwight may have been asked to deal with this material critically as a reader given his facility with the language.

After his time in Oxford, Dwight held an Ecclesiastical and Legal appointment to successive Bishops of Chester, until a deteriorating relationship with the fourth to hold the title led to him leaving the post to open the Fulham Pottery in 1670/1. During

[20] Haselgrove & Murray, "Dwight's Fulham Pottery," 22–8.

[21] Foster, *Alumni Oxonienses*; and Boyle, *Will,* 11/408/169.

[22] Reproduced in Haselgrove & Murray, "Dwight's Fulham Pottery," 142.

[23] Newman & Principe, *Alchemy Tried in the Fire,* 211–2.

this period between 1661 and 1670, no records connecting Dwight to chymistry or ceramics of any kind are extant. This lack of documentary evidence as to Dwight's endeavors suggests that he was able to spend very little time in these pursuits, as within the same period his marriage was brought to fruition and a plethora of documents connected to this and his legal post remain. Some progress must have been made, as he was able to secure a royal patent for the production of porcelain for the factory almost immediately in 1672 supported by Hooke and Boyle.[24]

Despite the support of the Crown in the production of his wares, and the endorsements of individuals such as Hooke, the absence of any porcelain produced at the Fulham factory frustrated scholarship on Dwight for the majority of the nineteenth and twentieth centuries turning the focus primarily to his stoneware production. Fortunately, the role of the factory in the history of English ceramics, and its continued operation until the twentieth century provided both the interest and possibility of substantial archaeological excavations undertaken by the Archaeological Section of the Fulham and Hammersmith Historical Society between 1971 and 1979.[25]

These excavations revealed substantial information about many of Dwight's working practices, the composition of both his finished and unfinished ceramics, and substantiated the claims that he had been actively working on producing porcelain in the Chinese style. Various strata were excavated, and unearthed material covered virtually the entire operation of the pottery throughout its existence. The earliest period of experimentation and production, covering the years 1671/2–74 was found "confined to a few Features in the south-east of the site (K15, K23, N1, and D3)" with only very fragmentary material found in other areas.[26]

In these areas of early experimentation, a substantial sample of material of porcelaneous nature was discovered, all of which was fragmentary. Of this material, Green notes that less than "One kilo in total weight was recovered but [represents] a large number of individual vessels." The dating is connected to its location, relative to other excavations, with important finds coming from the aforementioned south-east corner *circa* 1673 and a pair of soak ways that were filled *circa* 1676 (Fig. 1).[27]

These sherds represent a systematic, sustained, and varied approach towards porcelain experimentation, one that resulted in many failures, but there were other small successes towards a production of porcelain. They provide a wealth of technical evidence towards the experimentation conducted by Dwight during the early period of the pottery, informing us specifically on the composition of his "porcelane" bodies, attempts and decoration, problems in glazing, and finally a variety of firing issues.

[24] The patent was issued towards the protected manufacture of 'transparent Earthenware' and 'stone ware' by Charles the Second, on 17 April 1672, at Whitehall. See Charles the Second, *Transparent Earthenware*, P.R.O. C. 82 2425 Cal. S.P. Dom Entry Book 34, fol. 155.

[25] The findings and details were published in Green, *Excavations*, ch. I–V.

[26] Green, *Excavations*, 11.

[27] Green, *Excavations*, 65.

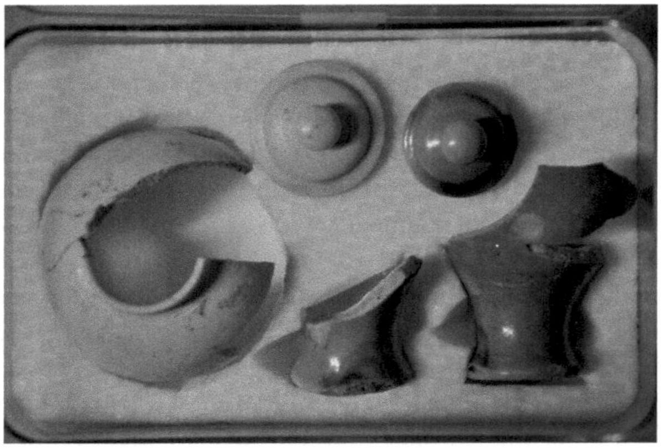

Fig. 1 Fragmentary porcelain material recovered at the Fulham site attributed to John Dwight, 1672–1673 (Photograph by the author)

Two groups of these sherds are particularly demonstrative. First, the chips intentionally produced to test a combination of bodies and surface solutions. These represent varying degrees of success, with a combination of the vapour glazes and dipped finishes (both slip and glaze) seen in the test pieces. Figure 2 illustrates pieces 33 (slip), and 34 (demonstrably dipped in a glaze, with crazing, and yellowing visible though with good transparency overall), and one of a number of sherds not illustrated in Green (vapour glazed).

The system for marking the test chips shows sophistication in Dwight's methodology, incorporating not only distinct numbers for various colors and finishes, but including descriptions of where each was applied. For example the fragment represented by type 36 clearly shows a division between the front and back of the test chip (Green speculates that it is divided as ‘$iq^{r[ecto]}/ iq^{v[erso]}$’), as do types 32, 34, and 48. It is a reasonable assumption, based on his later workbooks, that the various notations correspond to a notebook kept during the process of experimentation. This hypothesis is supported by the survival of two of Dwight's notebooks, containing various recipes and refinements. These two notebooks were of much later date than the excavated material, written between 1689 and 1695. It is likely that the latter was completed to hand down to his son Samuel Dwight, as an aid towards the operation of the factory. While Samuel's later ownership of the notebooks can only be speculated, the books themselves contained the signature of Lydia Dwight, claiming them as her own.

Modern knowledge of the two notebooks is due to the transcriptions by three individuals. The primary transcription was executed by Lady Charlotte Schreiber (1812–1895), who discovered these two books during a visit to the pottery in 1870, painstakingly recording their contents in her own notebook. Excerpts were also copied by William Chaffers (1811–1892) and Llewellyn Jewitt (1816–1886) for publication.[28] While these two transcriptions lack the recipes for porcelain recorded by Lady Schreiber, they corroborate the rest of her transcription.

[28] Chaffers, *Marks and Monograms*; and Jewitt, *Ceramic Art*.

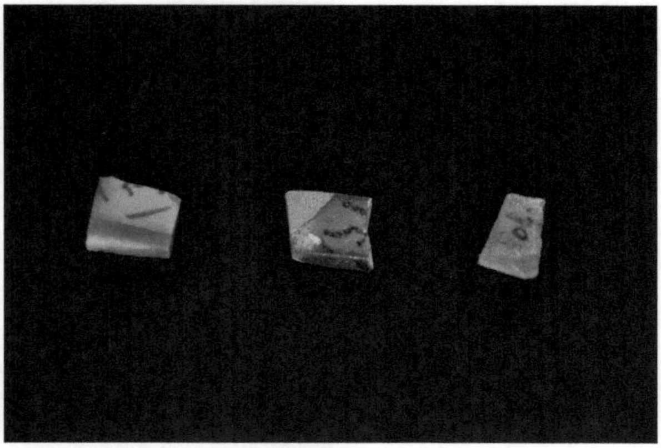

Fig. 2 Porcelain sherds, two identified as Green's type 33 and 34 the other unrecorded, attributed to John Dwight, 1672–1673 (Photograph by the author)

The current whereabouts of the two notebooks is unknown; they were sold through Christie's, likely in 1888 or 1889. Notices seeking their recovery have been remarked on as early as 30 June, 1894, but no indication as to their location has surfaced.[29]

While no contemporary records of the tests conducted between 1672 and 1675 are known to exist, laboratory testing of the composition of the sherds found from this period coincide broadly with the recipes given for porcelain in the later texts. This methodology demonstrates the connection between Dwight's laboratory practice to that employed by Boyle in the late 1650s, and echoes the laboratory books of Starkey.[30] Similar to the annotations and refinements demonstrated in Starkey's laboratory books, we see Dwight actively editing his working notebook with commentary on the effectiveness of recipes and the striking through of less effective formula. Lady Schreiber was rigorous in her transcriptions, making note of pages torn out, recipes struck through, and duplicating marginalia such as a "not very good" accompanying a struck through recipe for "a dark red porcellane or China Cley" dated 14 November, 1693.[31]

The second, arguably more interesting, group discovered within the context of soakway A18, and thus contemporary to the test chips, was a group of Chinese export porcelain sherds (Fig. 3). Among this group of blue and white sherds, representing at least five distinct objects, are cases where glaze running over fractured edges demonstrate clear evidence of re-firing, at a temperature hot enough to at least partially liquefy the glaze, without burning the cobalt. A final sherd from this group contains evidence of cobalt over-painting by Dwight, before re-firing, perhaps as a test coinciding with the samples of his own work.

[29] Schreiber, *Charlotte Schreiber's Notebook*, 31; and Haselgrove & Murray, "Dwight's Fulham Pottery," 74.

[30] For the working practices, see Starkey, *Alchemical Laboratory Notebooks*.

[31] Reproduced in Haselgrove & Murray, "Dwight's Fulham Pottery," 73, ai–bviii, 74.

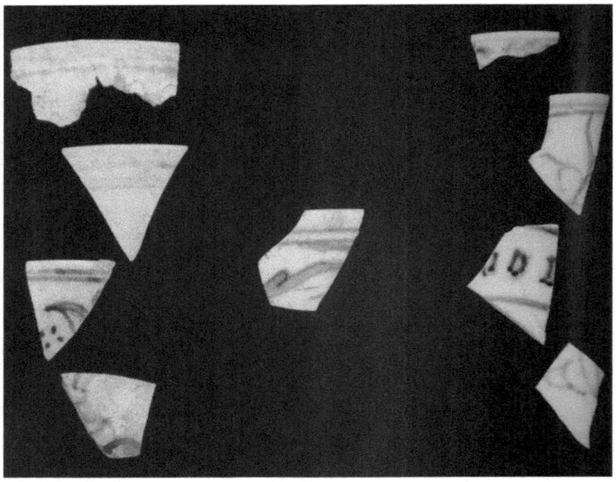

Fig. 3 Sherds from the period of 1671/2–1674 with evidence of re-firing, including five sherds of Chinese origin (*center* and *right*) (Photograph by Edwin Baker, Museum of London; Reproduced courtesy of English Heritage)

While the source of these sherds is unknown, whether they were Dwight's personal property, donations by an interested party, or the possessions of a funding patron, their role within Dwight's active search for the key to making porcelain is clear. These fragments provided a tangible outcome for Dwight's experimentation, sources that could be used for active comparison in the laboratory. His test bodies and colors worked directly towards the production of material that mimicked the observable qualities of the Chinese porcelain in his possession even though its chemical composition remained unknown.

Those qualities clarify his distinct use of "transparent porcelane or China cley" to describe certain recipes in his notebooks. While he employed "porcelane" as a wider term applicable to many finewares, it was transparency and whiteness that he pursued through experimentation, as key properties of the sherds available to him. The re-firing of Chinese material revealed the use of a distinct glaze layer over the cobalt itself, which demonstrated a distinct thermal response from the Chinese body paste. There may have been more sherds whose surfaces were destroyed by over-firing in a test furnace, but the remaining pieces show care taken to affect changes in the glaze without damaging the fabric or the decoration, suggesting that, on a smaller scale, Dwight had keen control over the temperatures of his experimentation. The bright, transparent nature of the Chinese glaze, combined with its excellent fit to the fabric of the ware, may have further influenced Dwight towards an essay discussing a theoretical lead China glaze, as it appears in his notebook, demonstrating that his attempts to create an alkali glaze had failed, or if successful, his solution remains one of the best kept secrets of European ceramics.

His application of thermal experimentation, as noted, was limited in its scope, but still formed a critical point of departure for his investigation into the properties necessary to replicate the Chinese materials. He combined these inquiries with knowledge of a wide range of raw materials available to potters in England.

In conversations recorded by Hooke in his diary, and referenced in at least one meeting of the Royal Society on 5th December 1678, Dwight was vocal about the suitability of various English clays for the production of "porcelane."[32] Further, Charles Leigh published a second hand account of the clays local to Wigan that were being employed by potters that Dwight "made his first Discovery" upon.[33] Dwight was thus actively pursuing a synthetic solution to the problems facing porcelain production using local materials in a methodical manner, the firing and re-firing of sherds being fundamental to his process.

Ehrenfried Walther Von Tschirnhaus: Harnessing the Sun

In contrast to the limited contemporary information regarding Dwight's activities and a lack of autograph material besides the two slender workbooks, an increasing amount of attention paid to Tschirnhaus since 1990 has provided numerous contemporaneous sources.[34] Tschirnhaus himself produced ample material as a long-standing contributor to the journal *Acta Eruditorum*, and various correspondences have survived.[35]

Historically, the focus on Tschirnhaus prior to the later half of the twentieth century focused on his contributions to Mathematics and his intellectual relationships with leading natural philosophers, including Leibniz, Spinoza, Boyle, and Athanasius Kircher (c.1601–1680). The historic minimization of his role in discovering the porcelain *arcanum* can be linked to the use of considered propaganda, as early as 1710–1720, in establishing a popular narrative around the Meissen Factory and Johann Friedrich Böttger (1682–1719), who was credited with the discovery after Tschirnhaus's death in 1708.

As archives have been fully absorbed into the framework of modern knowledge, this false historiography has eroded under the weight of documentary evidence. Based on the essays published in *Acta Eruditorum* between 1687 and 1697 and material found in the Dresden archives, the full importance of Tschirnhaus in establishing the foundations of Saxon porcelain is clear. These works confirm

32 Birch, *Royal Society*, vols. III–IV.

33 Leigh, *Natural History of Lancashire*, 56–7.

34 The biographical details of Tschirnhaus can be found in numerous sources, including Winter, "Tschirnhaus"; Watanabe-O'Kelly, *Court Culture*, both include excellent work on Tschirnhaus's biography and role within the industrial structure of Saxony. Further, the Staatliche Kunstammlungen Dresden Mathematisch-Physikalischer Salon, through the Tschirnhaus Gesellschaft has extensively assembled materials relating to his work and experiments.

35 The *Acta Eruditorum* represents what can be considered the first international science journal, published during the period between 1682 and 1782, as founded by Otto Mencke (1644–1707) and Gottfried Wilhelm Leibniz (1646–1716). Tschirnhaus was a frequent contributor to the journal, submitting no less than 15 letters or essays on mathematics, seven on burning lenses and one on astronomy over the course of 17 years.

that not only did the elder chymist act as the principal investigator into the *arcanum* from the 1690s until his death in 1708, but that the success of experimentation in Saxony hinged upon his refinement of the application of directed heat energy via burning lenses. Tschirnhaus's commitment to burning lenses can be traced to a meeting with François Villete (1621–1698) in Lyons in 1676, that marks the beginning of more than 20 years of focused experimentation involving them.[36]

The aim of this section is not to dive into the melee surrounding the credit for the final recipe for porcelain production as presented to Augustus the Strong in 1708–1709. However, the slow dissemination of Tschirnhaus scholarship outside Germany remains connected to that historiography, and so it must be touched upon. Ultimately the death of Tschirnhaus in 1708 combined with the period of refinement before the production of early commercial test wares is proof enough that Böttger was possessed of the skill set necessary to further the research. Rather, the focus must be on what contributions can be unquestionably linked to the work of Tschirnhaus, prior to Böttger's historic collision with the courts of Europe.[37]

Acta Eruditorum 1687–1695

The most direct evidence for Tschirnhaus's application of thermal analysis in the quest for porcelain is contained in a series of contributions he made to *Acta Eruditorum* explaining the construction and use of increasingly sophisticated burning lenses. By combining the manufacturing prowess of his native region with the fundamental principles communicated to him by Villette, Tschirnhaus was able to focus the sun's energy to achieve temperatures hitherto unknown in Europe. While many of his major findings were communicated directly to individuals across the continent, he felt that the discoveries were significant enough to submit for publication.

The first of these papers appears in the January 1687 issue of the journal and includes a cursory list of experimental observations of the reaction of a variety of metals, including lead, copper and gold, and related observations based on the time of focused exposure to the lens.[38] During this phase of experimentation, his test

[36] The most extensive discussion of these experiments can be found in Plassmeyer, *Sonnenfeuer*, which presents an extensive discussion of the wider reception of his endeavours, refinements connected to the burning lenses, and a complete catalogue of the lenses in the collections of Dresden.

[37] Pietsch, "Tschirnhaus," summarizes the current state of scholarship on the matter, including various visits and investigations by Tschirnhaus prior to the involvement of Böttger. It does however fail to discuss the comments made by Tschirnhaus in *Acta Eruditorum* during this period.

[38] Tschirnhaus, "Speculi ustorii."

objects included crucibles that would withstand the heat until past the eight-minute point.[39] Based on the times given for copper and gold before liquefying the data would suggest that these were likely the Hessian crucibles mentioned above.[40]

The following three works to appear in the pages of *Acta,* in April 1688, November 1691, and August 1696 respectively were significant contributions to the knowledge of thermal properties of a range of substances, but reveal no distinct experimentation on ceramic materials.[41] The essay from April of 1688 records the successful use of a burning mirror towards the liquefaction of asbestos. He goes on further to discuss the effect of the time of year on the power of the lenses, specifying that material that would melt in 8 or 9 min in the summer would take as long as 12 min in the frigid cold of January.[42]

The extended essay of November of 1691 included detailed instructions on the construction of the lenses with an accompanying figure presenting a schematic for the angles involved in correctly aligning their focus. Tschirnhaus presents an extended discussion of experimental results in this volume, before posing larger philosophical questions.

A period of 5 years separates the appearance of the schematics in the journal until publication of technical refinements and corrections in August of 1696. Tschirnhaus then followed quickly with a list of results in September of 1697. It is in this, the last of his works published on burning lenses that the direction of his research towards ceramics specifically appears.[43] While correspondence with Leibniz suggests that Tschirnhaus could claim a limited success in the production of porcelain as early as 1694, no material evidence has appeared.[44] However, in the pages of *Acta*, he includes the results of the application of burning lenses on combinations of material with *porcellana Hollandica* (Dutch earthenware) and *porcellana Chinensi.*

Tschirnhaus observed that the material combined with *porcellana Hollandica* burst immediately into flames at the temperature of experiment, while that made with Chinese porcelain formed spherical glass and asbestos became entirely translucent.[45]

When measured against the long list of other substances included in the 10 years of published experiments, it is evident that Tschirnhaus had a large amount of relative thermal data to use as a metric for his porcelain experiments. While other

[39] Tschirnhaus, "Speculi ustorii," 53.

[40] Martinón-Torres, Rehren & Freestone, "Mystery of Hessian Wares."

[41] Tschirnhaus, "Paralipomenon," "Singularia effecta," and "Artis vitriariae."

[42] Tschirnhaus, "Speculi ustorii," 52.

[43] Tschirnhaus, "De magnis lentibus."

[44] Reinhardt, "Ehrenfried Walther von Tschirnhaus," 13, and "Tschirnhaus oder Böttger?," 32.

[45] "4. Lateres, lapis scissilis, pumex, porcellana Hollandica, asbestus, cujuscunque sint magnitudinis, statim ignescunt, & facile in vitrum convertuntur [...] 8. Si fragmina minora ex lateribus, lapide scissili, porcellana Chinensi, talco &c. carboni tali imponantur, momentum funduntur, & in globulos abeunt vitreos. Asbestus totus in pellucidum vitreum globulum convertitur." (Tschirnhaus, "De magnis lentibus," 415–6).

authors have commented upon the possible influence of Hessian crucibles on the discovery of porcelain in Saxony, the geographic separation of the two regions argues against a direct link.[46] However, the data from January 1687 regarding the heating of crucibles when combined with the later investigations on earthenware and porcelain would have been a clear indicator that a search must be undertaken for Saxon clay with the same refractory properties as the crucibles. The clay previously employed for production of earthenwares was clearly inferior within the context of the experiments of 1696–1697.

The other European porcelains provided a robust collection of comparative samples, particular in the context of body fabrics that would lack the thermal tolerances of Chinese porcelain. With the experimental evidence supplied by the trials published in *Acta*, Tschirnhaus's later journey to Delft and St. Cloud in 1702–1703 was principally an opportunity to refute the local factories' techniques. His correspondence bears this out, as he was able to readily discard the possibility of their productions being true porcelain when communicating his thoughts back to Saxony.[47] While later work on the porcelain *arcanum* was divided between Tschirnhaus and Böttger, based on the material characteristics of early Meissen porcelain it is undeniable that the unique approach to thermal experimentation with burning lenses was critical in achieving the first successes.

The Legacy of Thermal Analysis on European Porcelain Production

Previous scholarship has rightly raised the issue of missionary and trade records that contained practical information on the Chinese production techniques.[48] Certainly records as early as the writings of Gaspar da Cruz (c.1520–1570), in his *Tractado em que se cotam muito por esteso as cousas da China* of 1569, offered an accurate description of porcelain's constituents of clay and white soft stone.[49] No direct links between these texts and the work of the seventeenth- century *arcanists* can be drawn, or even with the influence of textual sources on the ceramicists of the eighteenth century. What remains in the historic record is unequivocal: even with these texts in circulation for at least a century, no successful production of commercial porcelain was realized in Europe, despite the availability of kaolinized clay sources in both Saxony and England. I argue that this fact alone affirms that a critical component was lacking within the knowledge infrastructure of the

[46] Zumbulyadis, "Böttger's Eureka!," tries to tie the invention exclusively to Böttger's exposure to crucibles, unsuccessfully, but raises excellent questions regarding the relationship of other German high fire wares with Meissen porcelain.

[47] Reinhardt, "Tschirnhaus oder Böttger," 39, 43.

[48] Pietsch, "Tschirnhaus."

[49] Cruz, *Tractado*.

workshop. While an eventual solution may have been found, the possibility was circumvented by the engagement of the chymists with the problem of porcelain production.

It is clear that both Dwight and Tschirnhaus were concerned with asking similar questions of Chinese porcelain and their results directed the future avenues of their experiments. As previously mentioned, both sets of data were limited in their scope and restricted to comparative discourse with local material. The efficacy of analytic techniques on ceramic bodies can also be seen to play a role, with the slow firing of sherds in Dwight's various furnaces or kilns restricting his ability to both observe the objects as they processed through the sequence and the time frame for his analysis. If one considers the heating rate of a seventeenth-century stoneware kiln, or even a smaller furnace, its maximum achievable temperature and the amount of energy necessary to vitrify or melt a sample of porcelain, the process would have been fuel intensive and taken hours per test, with little direct control of the experimental environment. Tschirnhaus's burning lenses allowed for results to be obtained in minutes, and greater specificity in the temperature and period of heating.

The limitations placed on Dwight proved to be insurmountable to the production of a commercially viable porcelain body, yet resulted in substantial progress made towards that end goal. His synthetic additive allowed the creation of material that was chemically similar to some of the porcelains being produced in China, his ultimate failure resting on his inability to solve the issues of glazing and material stability while firing. That pivotal problem accounts for the excavated sherds that show re-firing at temperatures hot enough to affect the glaze, but not the body. Tschirnhaus, through the eventual completion of his work by Böttger, was able to achieve the basis for a fully vitrified, translucent, white body. An examination of the elemental composition of the early 'Böttger' wares, produced in the original phase of production, reveals that while they share the observable physical qualities of Chinese porcelain, they must be viewed as an entirely separate technical solution. This separation of property from composition demonstrates both the strengths of Tschirnhaus's and Böttger's approach in reproducing observable qualities, and the inaccessibility of the chemical fundamentals of the Chinese material.

A comparison of modern elemental analyses of 'Dwight' experimental ware, *Blanc-de Chine*, 'Böttger' ware, and *Kangxi* porcelain reveals these deep divisions in their compositions, while all meet similar physical criteria (Table 1). The last line of the table is an analysis of the body composition of porcelain produced at Meissen after approximately 1727 demonstrating changes that reduced the firing temperature of the material and brought it more directly into line with its Chinese inspiration. That this change was made after the gradual dissemination of the letters of Père d'Entrecolles describing the process observed in China must be viewed as a final indication of the impact that thermal experimentation had on establishing the qualities of the earliest commercial European porcelain.

The chemical differences are a quantifiable demonstration that, while allowing a sophisticated application of experimental process to solve the problem of producing porcelain, thermal experiments in the seventeenth century were unable to fully

Table 1 A comparison of modern elemental analyses of 'Dwight' experimental ware, *Blanc-de-Chine*, 'Böttger' ware, and *Kangxi* porcelain

Material	SiO_2	Al_2O_3	Fe_2O_3	CaO	MgO	K_2O	Na_2O	Other	Total
Dwight	80.3	12.4	0.74	0.39	0.2	5.35	0.47	Trace	99.1
Blanc-de-Chine (ft46)	76.7	16.8	0.35	0.15	0.08	5.9	0.2	Trace	100
Kangxi Porcelain (ft47) seventeenth century	64.7	28.35	0.95	0.5	0.1	2.8	2.4	0.19	99.99
Böttger Porcelain 1708	61.0	33.0	0.00	4.8	0.00	0.1	0.2	0.9	100
Meissen Porcelain (ft48) mid-eighteenth century	59.0	35.00	0.00	0.3	0.00	4.0	0.00	0.9	100

answer the questions put to them. The results allowed innovative solutions to the problem of a functional translucent body, while at the same time both Dwight and Tschirnhaus were diverted from a solution that would duplicate the materiality of the Chinese material.

Chymical Laboratory as Potter's Workshop

The early modern inclusion of ceramics within the arts of the fire brought with it access to the experimental techniques developed by alchemist and chymists. This shift allowed an interrogative approach to the production of porcelain that utilized many of the tools traditionally found in the craftsman's workshop while circumventing the gradual processes of craft innovation. When we consider chymistry as both a kind of knowledge, and an object-producing discipline, the application of its techniques to reverse the process of making porcelain is a logical extension of the economic demand for china and the intellectual drive toward maker's knowledge as expressed by Bacon.

Certainly, the history of ceramics after the seventeenth century is inextricably linked to chemistry's success at solving increasingly sophisticated questions of luxury pottery production. We need look no further than William Cookworthy's (1705–1780) patent for the manufacture of porcelain in 1768 or the thousands of experiments conducted by Josiah Wedgwood (1730–1795) in the production of his Jasperwares. Unlike the other arts of the fire; metalworking, glassmaking, and dyeing, the late inclusion of ceramics into the hierarchy of decorative arts allows contemporary scholars opportunities to investigate the separations between the craft artisan and the chymist during the early modern period, of which this discussion of thermal analysis is only one of many.

Bibliography

Bacon, Francis. 1858. *New Organon*, trans. James Spedding et al. London: Longmans and Co.

Birch, Thomas. 1756–1757. *History of the Royal Society of London*. London: A. Millar.

Biringuccio, Vannoccio. [1540] 1942. *The Pirotechnia of Vannoccio Biriniguccio: The Classic Sixteenth-Century Treatise on Metals and Metallurgy*, eds. and trans. Cyril Stanley Smith and Marta Teach Gnudi. New York: Dover Publications.

Boardman, John. 1988. Trade in Greek Decorated Pottery. *Oxford Journal of Archaeology* 7: 27–33.

Boyle, Robert. 1661. *The Sceptical Chymist or Chymico-Physical Doubts and Paradoxes, Touching the Spagyrists Principles Commonly call'd Hypostatical: As They Are Wont To Be Propos'd and Defended by the Generality of Alchymists. Whereunto is Premis'd Part of Another Discourse Relating To the Same Subject*. London: J. Cadwell for Crooke.

Boyle, Robert. 1691. *Will of the Hon. Robert Boyle, F.R.S, July 1691*. The National Archives, Manuscript PROB 11/408/169.

Carswell, John. 2000. *Blue and White: Chinese Porcelain Around the World*. London: British Museum Press.

Chaffers, William. 1870. *Marks and Monograms on Pottery and Porcelain*, 3rd ed. London: J Davy and Sons.

Charles the Second. 1672. *Patent for the Protected Manufacture of Transparent Earthenware*. Public Record Office. C. 82 2425 Cal. S. P. Dom Entry Book 34.

Cruz, Gaspar da. 1569. *Tractado em que se cotam muito por esteso as cousas da China*. Madrid: em casa de Andre de Burgos.

Debus, Allen G. 1967. Fire Analysis and the Elements in the 16th and 17th Centuries. *Annals of Science* 23: 127–147.

d'Entrecolles, Père Francois Xavier. 1781. Lettre D'Entrecolles à Jao-tcheou, 1er Septembre 1712' and 'Lettre D'Entrecolles à Kim-te-tchim, le 25 Janvier 1722. In *Lettres édifiantes et curieuses écrites des missions estranges. Mémoires de la Chine etc.*, vol. 18–19. Paris: Societas Jesu Missio Sinensis.

Foster, Joseph. 1891. *Alumni Oxonienses 1500–1714*. London: Parker and Co.

Green, Chris. 1999. *John Dwight's Fulham Pottery: Excavations 1971–1979*. London: English Heritage.

Haselgrove, Dennis, and John Murray. 1979. John Dwight's Fulham Pottery 1672–1978: A Collection of Documentary Sources. *Journal of Ceramic History* 11: 1–284.

Jewitt, Llewellyn. 1878. *The Ceramic Art of Great Britain*. London: Virtue and Co.

Kerr, Rose, and Nigel Wood. 2004. *Ceramic Technology*. In *Science and Civilization in China*, vol. 5, ed. Joseph Needham. Cambridge: Cambridge University Press.

Kingery, William David, and Pamela B. Vandiver. 1984. Medici Porcelain. *Faenza* LXX(5–6): 441–453.

Kingery, William David, and Pamela B. Vandiver. 1986. *Ceramic Masterpieces*. New York: Free Press.

Leigh, Charles. 1700. *The Natural History of Lancashire, Chesire and the Peak in Derbyshire with an Account of the British Phoenician, Armenian, Greek and Roman Antiquities in Those Parts*. Oxford: George West and Henry Clement.

Liverani, Giuseppe. 1936. *Catalogo delle Porcellane dei Medici*. Piccola Biblioteca del Museo delle Ceramiche. *Faenza* 2.

Martinón-Torres, Marcos, Thilo Rehren, and Ian Freestone. 2006. Mullite and the Mystery of Hessian Wares. *Nature* 444: 437–438.

Newman, William R. 1994. *Gehennical Fire: The Lives of George Starkey an American Alchemist in the Scientific Revolution*. Cambridge, MA: Harvard University Press.

Newman, William R. 2004. *Promethean Ambitions: Alchemy and the Quest to Perfect Nature*. Chicago: University of Chicago Press.

Newman, William R., and Lawrence M. Principe. 1998. Alchemy vs. Chemistry: The Etymological Origins of a Historiographic Mistake. *Early Science and Medicine* 3: 32–65.
Newman, William R., and Lawrence M. Principe. 2002. *Alchemy Tried in the Fire: Starkey, Boyle and the Fate of Helmontian Chymistry*. Chicago: University of Chicago Press.
Pietsch, Ulrich. 2001. Tschirnhaus und das europäische Porzellan. In *Ehrenfried Walther von Tschirnhaus (1651–1708): Experimente mit dem Sonnenfeuer*, eds. P. Plassmeyer and S. Siebel, 68–74. Dresden: Staatliche Kunstammlungen.
Plassmeyer, Peter. 2001. *Ehrenfried Walther von Tschirnhaus (1651–1708): Experimente mit dem Sonnenfeuer*. Dresden: Staatlich Kunstammlungen Dresden.
Reinhardt, Kurt. 1903. Beiträge zur Lebensgeschichte von Ehrenfried Walther von Tschirnhaus. In *Jahresbericht der Fürstenund und Landesschule St. Afra in Meissen*. Meissen: Klinkicht & Sohn.
Reinhardt, Kurt. 1912. Tschirnhaus oder Böttger? Eine urkundliche Geschichte der Erfindung des Meissener Porzellans. *Neues Lausitzisches Magazine* 88: 1–162.
Starkey, George. 2004. *Alchemical Laboratory Notebooks and Correspondence*, eds. and trans. William R. Newman and Lawrence M. Principe. Chicago: University of Chicago Press.
Schreiber, Lady Charlotte. 1874. *Lady Charlotte Schreiber's Notebook*, Manuscript, British Museum OA.6557.
Theophilus, Presbyter. 1847. *An Essay Upon Various Arts in Three Books, by Theophilus, Called also Rugerus, Priest and Monk, Forming an Encyclopedia of Christian Art of the Eleventh Century,* ed. and trans. Robert Hendrie. London: John Murray.
Theophilus, Presbyter. 1961. *Theophilus, the Various Arts: De Diversis Artibus*, ed. and trans. C. R. Dodwell. Oxford: Clarendon.
Theophilus, Presbyter. 1963. *Theophilus On Diverse Arts: The Foremost Medieval Treatise on Painting, Glassmaking and Metalwork*, eds. and trans. John G. Hawthorn, and Cyril Stanley Smith. New York: Dover.
Tschirnhaus, Ehrenfried Walther von. 1687. Relatio de insignibus novi cujusdam speculi ustorii effectibus. *Acta Eruditorum* 6: 52–54.
Tschirnhaus, Ehrenfried Walther von. 1688. Paralipomenon ad Acta mensis Januarii 1687, pag. 52 de specula cujusdam ustorii singularibus effectibus. *Acta Eruditorum*: 206.
Tschirnhaus, Ehrenfried Walther von. 1691. Singularia effecta vitri caustici bipedalis, quod omnia magno sumto hactenus constructa specula ustoria virtute superat. *Acta Eruditorum*: 517–520.
Tschirnhaus, Ehrenfried Walther von. 1696. Intimatio singularis novaeque emendationis Artis vitriariae. *Acta Eruditorum*: 345–347.
Tschirnhaus, Ehrenfried Walther von. 1697. De magnis lentibus seu vitris causticis, quorum diameter trium quatuorve pedum, nec non euorundem usu & effectu plene & perspicue indicato. *Acta Eruditorum*: 414–419.
Tschirnhaus, Ehrenfried Walther von. 2000. *Gesamtausgabe. Reihe II, Amtliche Schriften*, ed. Eberhard Knobloch. Leipzig: Verlag der Sächsischen Akademie der Wissenschaften.
Watanabe-O'Kelly, Helen. 2002. *Court Culture in Dresden: From Renaissance to Baroque*. Basingstoke: Palgrave Macmillan.
Wilson, Timothy. 2007. Le maioliche. In *Il Rinascimento Italiano e l'Europa*. Vol. IV, *Commercio e cultura mercantile*, eds. Franco Franceschi, Richard A. Goldwaithe, and Reinhold C. Mueller, 217–245. Treviso: Fondazione Cassamarca; Costabissara (Vincenza): Angelo Colla.
Winter, Eduard. 1960. Der Bahnbrecher der deutschen Frühaufklärung. E.W.v. Tschirnhaus und die Frühaufklärung in Mittel- und Osteuropa. In *Quellen und Studien zur Geschichte Osteuropas*, III. Berlin: Akademie-Verlag.
Zumbulyadis, Nicholas. 2010. Böttger's Eureka!: New Insights into the European Reinvention of Porcelain. *Bulletin for the History of Chemistry* 35: 24–32.

MIX
Papier aus verantwortungsvollen Quellen
Paper from responsible sources
FSC® C105338
FSC
www.fsc.org

If you have any concerns about our products,
you can contact us on
ProductSafety@springernature.com

In case Publisher is established outside the EU,
the EU authorized representative is:
Springer Nature Customer Service Center GmbH
Europaplatz 3, 69115 Heidelberg, Germany

Printed by Libri Plureos GmbH
in Hamburg, Germany